Cunning Machines

Chapman & Hall/CRC Artificial Intelligence and Robotics Series

Series Editor: Roman Yampolskiy

Intelligent Autonomy of UAVs
Advanced Missions and Future Use
Yasmina Bestaoui Sebbane

Artificial Intelligence
With an Introduction to Machine Learning, Second Edition
Richard E. Neapolitan, Xia Jiang

Artificial Intelligence and the Two Singularities
Calum Chace

Behavior Trees in Robotics and AI
An Introduction
Michele Collendanchise, Petter Ögren

Artificial Intelligence Safety and Security
Roman V. Yampolskiy

Artificial Intelligence for Autonomous Networks
Mazin Gilbert

Virtual Humans
David Burden, Maggi Savin-Baden

Deep Neural Networks
WASD Neuronet Models, Algorithms, and Applications
Yunong Zhang, Dechao Chen, Chengxu Ye

Introduction to Self-Driving Vehicle Technology
Hanky Sjafrie

Digital Afterlife
Death Matters in a Digital Age
Maggi Savin-Baden, Victoria Mason-Robbie

Multi-UAV Planning and Task Allocation
Yasmina Bestaoui Sebbane

Cunning Machines
Your Pocket Guide to the World of Artificial Intelligence
Jędrzej Osiński

For more information about this series please visit:
https://www.crcpress.com/Chapman--HallCRC-Artificial-Intelligence-and-Robotics-Series/book-series/ARTILRO

Cunning Machines

Your Pocket Guide to the World of Artificial Intelligence

Jędrzej Osiński

CRC Press
Taylor & Francis Group
Boca Raton London New York

CRC Press is an imprint of the
Taylor & Francis Group, an **informa** business

A CHAPMAN & HALL BOOK

First edition published 2020
by CRC Press
6000 Broken Sound Parkway NW, Suite 300, Boca Raton, FL 33487-2742

and by CRC Press
2 Park Square, Milton Park, Abingdon, Oxon, OX14 4RN

ISBN: 978-0-367-89861-8 (hbk)
ISBN: 978-0-367-89802-1 (pbk)
ISBN: 978-1-003-02153-7 (ebk)

Typeset in Minion

by Deanta Global Publishing Services, Chennai, India

To my lovely wife Monika,
Who is so charming, spontaneous and full of empathy
that could never be replaced by AI,

And to my wonderful son Kacper,
Whose never-ending questions
keep my neurons active seven days a week.

"If you can't explain something simply,
you don't understand it well enough."

ALBERT EINSTEIN

Contents

Scientific Manifesto, xi

Author, xiii

CHAPTER 1 ▪ Introduction: Magic Revealed   1

CHAPTER 2 ▪ Artificial Intelligence: A Sci-Fi Phrase That
            Changed the World   7

  WHAT IS AI?   7

  STRONG ARTIFICIAL INTELLIGENCE   8

  TURING TEST   12

  ALGORITHM OR HEURISTIC?   14

  "I WANT TO PLAY A GAME"   25

  ARE WE THERE YET?   35

CHAPTER 3 ▪ Neural Networks – A Brainstorm inside a PC   41

  EVERYTHING IS A NUMBER   45

  SECRETS OF ARTIFICIAL BRAINS   49

  BRAINSTORMING   57

  LAYERS AND LIARS   61

  ARTIFICIAL REASONING   64

  DEEP THOUGHT   74

CHAPTER 4 ∎ Genetic Algorithms: From the Galapagos
Islands to a Computer-Composed Symphony   81

EVOLUTION SQUEEZED INTO MILLISECONDS   83

ARTIFICIAL DNA   86

THE BIRTH OF LIFE   88

NATURAL SELECTION   89

CROSSOVER: A NEW GENERATION BUILDS A NEW
WORLD   91

X-MEN AMONG US   96

EVOLUTION OF A SOLUTION   97

EVOLUTION IN IT   101

CHAPTER 5 ∎ Monte Carlo Method: An Unexpected Benefit
of Gambling   105

HOW MUCH IS Π?   112

CHAPTER 6 ∎ Language Processing: Plato and Expert
Systems   121

SYNTAX: PLAYING WITH BRICKS   126

FROM WORDS TO READING BETWEEN THE LINES   128

DATING ROBOTS   134

CHAPTER 7 ∎ A Future with Artificial Intelligence   139

BLACK BOX   141

POSTPONE YOUR UNEMPLOYMENT   145

GUNS AND ROSES   150

TABULA RASA   159

CHAPTER 8 ∎ Final Thoughts   163

INDEX, 167

Scientific Manifesto

B EFORE GOING ANY FURTHER, here is the one sentence I truly believe in and I want to ask you to keep in your mind from now on: everyone can be a scientist. I would say, even stronger: everyone is a scientist.

Looking back at the history of our civilisation, we sooner or later realise that nobody has ever been born or educated to be a world-changing scientist or inventor. Pierre de Fermat, a seventeenth-century genius whose results have been influencing mathematicians for generations, was earning his living as a lawyer. Albert Einstein revolutionised the whole field of physics while working as a clerk in a patent office.

The problem is that nowadays people are discouraged from science. When Einstein's theories were experimentally confirmed, he became a celebrity more recognised than music stars. World-renowned newspapers announced contests to describe the theory of relativity in the simplest possible form. And there were thousands of participants – people of many different professions, not only physicists. The world was truly interested in the topic. Today there are more discoveries made every year than there were during whole decades in the previous century. But we do not see them. Or do not want to see. Or do not have time for it. Why?

The first reason is the rush. Being in a hurry is simply trendy. Unfortunately, the same happens in school and in the workplace. I have heard it many times as an academic teacher: "I won't need that. I just want to learn the things I will use at work". And that is the beginning of the end. The end of creativity. The knowledge people which are taught is more and more limited, focusing only on specific tasks. From bright, thinking beings we are slowly changed into tools. People start to avoid theories and learn only their applications. There is no time for understanding the concept – we have just enough to employ the solution we have been shown. Having no time to ask *why*, we are left only with *how*.

The second reason is one of the biggest lies being successfully repeated everywhere to younger and younger people. They are told they will be unable to understand something. Some are said to be *humanists*, so they should not even try to look at a mathematical equation (as they could go blind from it). People are somehow classified after the first attempt at something. Did you know that Einstein's first version of his PhD dissertation was rejected? Or that for over two years he was unsuccessfully looking for employment as a teacher? What would he be told nowadays? Probably something like: "Leave it. Just learn to use this software. Companies need people that know that. And they pay well".

Do not give up knowledge. Do not look at visible benefits. Do not resign if your boss tells you it is not required. Contact with science does not only develop your skills but, much more importantly, it develops your mind. If you teach your brain to avoid challenges, you will sooner or later stand in front of a situation in which you have absolutely no idea what to do. A void in your head. Simply teach yourself to be interested in the world. Leave your work procedures and standards for a second and look at your work as a spectator. Stand for a moment on a street and think about the material it is made of.

And the great thing is you don't need to learn to be creative. You just need to remind yourself of how it is to be creative. Kids are very creative and ask thousands of questions. It is just the case that as they grow older, they are told that they will not need all the answers.

Never stop asking questions. Curiosity is one of our strongest instincts. Do not fight it, treat it a gift. Learn and be proud of it, despite your age. A scientific view may help you to find solutions at work and in your personal life. Curiosity keeps your mind active and prevents dementia.

Do not believe anybody who says you are unable to understand something. We are all born with the same curiosity in our minds and hearts. Follow it. Everyone is a scientist.

Author

Jędrzej Osiński, PhD, earned his PhD in artificial intelligence and has a strong research background in computer science. He has published 14 scientific papers to date, as well as co-authoring two books. During his academic career, he has worked on government grants and lectured on software testing and products, AI applications and IT business start-ups. He has participated in a number of different conferences as a speaker, paper reviewer and organiser. His scientific interests are mostly focused on spatiotemporal modelling and reasoning, knowledge representation, natural language processing, qualitative calculus and fuzzy sets.

Dr Osiński has over ten years of experience working in IT companies of different sizes, domains (the web, telecoms, banking, e-learning), organisation structures and locations (Poland, Ireland and the UK). He currently combines the roles of certified senior QA engineer, line manager and business analyst to support software projects for some of the world's best-known brands.

He is also involved in various initiatives promoting AI, science and modern technologies, including blog posts, science days and invited talks, as well as radio and TV appearances.

CHAPTER 1

Introduction: Magic Revealed

A RTIFICIAL INTELLIGENCE (AI) ARE two words heard not only in cinema movies and rocket scientists' offices, but more and more often repeated in the spaces free from these words just a few years ago. Nowadays, the phrase is being said by people of business, heads of churches, world-leading politicians, and it is not part of any joke – it appears in the same sentence as words like *future*, *civilisation*, *hope* or *danger*. The truth is that, although it is an expression repeated by many, it is well understood by just a few.

This is why AI systems are a little bit like magicians' tricks. We love to watch magic and, although we all know it is just an illusion, we do not want it to stop. We want to dive into it and feel like children once more. This book explains artificial intelligence. I will help you to understand what it really means, how it works and what we can expect from AI. The magic will be revealed, the dense smoke will dissipate and you will be allowed to enter to the magicians' backstage. That is why I give you here my **first warning**: make sure you want to learn all of this, as the next time you hear *artificial intelligence* mentioned on TV, you will not be as thrilled and excited as before. So now is your last chance to stay in the audience – close this book, put it on the shelf, leave the store and never come back. But if you decide to proceed, I can promise you things even more fascinating: the true magic of AI, like all of science, is that the more you know, the more you want to learn. Knowledge is addictive. And that is the **second warning**. Think twice...

Still here? I am very happy about that and I think you have chosen wisely. Let me guide you through the amazing world of artificial intelligence...

There are many topics surrounding artificial intelligence that we will discuss later on in this book; however, there is one that is so important that I would like to write about it here, before our further journey begins. The topic is feelings. There are a lot of them around AI, which can be noticed in the public space. One of the strongest one is **fear**. Fear, about ourselves, our families and our future. Fear, which is smartly used by the entire science-fiction entertainment industry. So, are our nightmares reasonable?

First of all, fear is pretty natural when we encounter something unknown, things we have never seen before or processes that we cannot explain. There is no reason to be ashamed of it – it is our evolution-made instinct used to help our ancestors to survive in a wild. Similar fears to today could be seen in the 1890s when first cinemas were constructed: people were so terrified of a train approaching (on the screen) that examples of audience panic escapes were recorded during early premieres. Nowadays, however, even 3D movie producers need to work pretty hard to get our heart rate up.

It is also important to understand that some risks may be mitigated just by the way we will actually use artificial intelligence in the future. Looking for an analogy, we can say it is a little bit like working with high-voltage electricity: if you do not have enough knowledge and protective clothing, you may be killed. But it does not mean that *the electricity* does it intentionally. Similarly, even if AI becomes self-aware and creative, it is our responsibility to teach it what is good and what is bad. It would be an empty sheet at the very beginning with no inherited reflexes or instincts (which can be encountered in the case of animals – a dog may be extremely dangerous sometimes despite good upbringing from a puppy). So it is mostly up to humans to define well the lines which are not allowed to be crossed. We will be the creators – likely and unfortunately.

I have written this book with one main purpose – to let everyone learn what artificial intelligence is and how it works. I have been active within that branch of science for the last 12 years, completing both an MSc and a PhD in these topics, working within government grant projects, delivering university lectures, speaking at conferences and contributing to the popularisation of AI science. What I have been noticing constantly is that AI is very often not understood correctly by people who speak about it in mass media. On the other hand, specialists usually focus on deep and advanced technical details that are almost impossible to understand by someone without a similar background. All of these factors generate more myths and mistakes, which makes the whole topic even more complicated and confusing.

But the truth is that the foundations of artificial intelligence are not rocket science. You do not need a PhD to understand how a basic neural network works. In fact, you do not even need advanced computer skills to learn it. Of course, both of the above may be required if you want to dive deeply into AI to implement your own solution that could be shared or sold. But as long as you simply want to understand the foundations, no extra preparation is required. If someone says you are not ready for that kind of knowledge, it means that the person wants to sell you an expensive preliminary course or just does not want to take the time to explain. Or, simply, that they do not understand it well enough since they are using one trained solution, repeated at work without any reflection. This book takes a different approach. Desire and curiosity are the only prerequisites I expect from you. No background is needed. You can switch off your PC as well (unless you know software programming at some middle level – see later). You may just sometimes need paper and a pencil.

As said already, this book is for everyone – even if you are not a tech person and do not see yourself working with AI a lot in future. Besides the concepts of basic artificial intelligence solutions, I will also relate their history and the inspirations that made them materialise – stories of unexpected connections between the world we see every day and high science breakthroughs. I believe this can also change your own way of perceiving the world. I hope you will find inspiration not only to learn more about AI but also to ask more questions and to look for your own solutions rather than simply follow standards (whatever your job is). That is how the biggest inventions started – from a question and the will of change. Be open-minded, ask questions, and train your curiosity every day.

I think this book may be also interesting for software developers and architects who work with or want to work with AI. Even if you use the algorithms at work, how well do you understand their background? Are you able to explain your results well to your non-tech boss, business department or a client? This book may help. Maybe you are a teacher looking for some new class exercises or easy metaphors? You should read it too.

So, enough of the introduction. Now a few words about the structure of the book itself. You are just a few lines away from completing this welcome chapter. The next one (Chapter 2) explains the main concepts, what does artificial intelligence really mean, where to find it, how scientists try to evaluate it and what are its main limitations. Chapters 3–6 describe the most popular artificial intelligence techniques in an easy form, together with background, anecdotes and some simple examples that not only help you to feel comfortable in such topics but give a little bit of fun and entertainment as well. The explained areas of AI in this book are

- Artificial neural networks
- Genetic algorithms
- The Monte Carlo method
- Natural language processing
- Ontologies and their applications

Chapters 3–6 may be read in random order; however, I still suggest following the original sequence. In Chapter 7, we will discuss the future of AI, what we can expect in a few years and in a few decades and what are the opportunities and risks of AI. Last but not least is the acknowledgements – there are many people I would like to thank. Without their help and support, you would never have had a chance to read this book.

There are two special editorial notations that I use to make it easier to navigate through the book, highlight extra content and better summarise the knowledge learnt. The first notation is a frame with a rocket icon (🚀) at the top – I use these frames to mark some additional piece of text related to a topic currently discussed but being a little bit more advanced (and thus may require some math or tech background). Of course, this is not true rocket science at all; however, you can skip these parts without losing further context. You can also return to these sections later while reading a chapter again. Treat them as pointers to further personal research if you find a specific aspect more interesting. Feel free to decide how far to follow it. The second notation is a summary frame with a pencil icon (✎) that you can find at the end of each book chapter. These frames contain a list of brief notes that recall the most important concepts and thoughts of a specific chapter. You will also find the index at the very end of the book, I have prepared for everyone who wishes to quickly go back to the most interesting concepts or to check whether a specific topic is discussed in the volume.

The topic of AI is equally important and interesting to IT developers as well as to non-technical people. The future is considered not only by world leaders but is also a breakfast topic of average families. All people are born to be scientists; it is just that the environment and everyday challenges do not always give us enough time to look at the sky...

So that is it. No more introduction. No more boringness. It is time for the true entertainment and the close scrutiny of the joy of science. You will soon learn things understood by few. Enjoy this, and feel proud of yourself. Open your mind, take a deep breath and dive into the world of artificial intelligence. You are entering the backstage area of today's biggest magic show. All secrets will soon be fully revealed...

✎ NOTES

- Artificial intelligence, although a phrase widely repeated in mass media, is not usually well understood by people who talk about it a lot.
- Fear of the new is a natural human reaction. It is not something you should be ashamed of. But it is worth studying the causes of your fears.
- Artificial intelligence is neither good nor bad. It has no personality; it is a tool. We can harm ourselves with it in the same way as we can be hurt by a hammer.
- Do not believe anyone who says some topic is outside your understanding. That person does not know it well enough.
- Open yourself to curiosity. Train it and care about this part of your personality, and you will find unexpected solutions for your everyday issues.
- Enjoy this book!

✎ YOUR NOTES

CHAPTER 2

Artificial Intelligence: A Sci-Fi Phrase That Changed the World

A RTIFICIAL INTELLIGENCE IS A buzz phrase nowadays. It truly is. Type it into Google, turn on the TV, read the latest newspaper, or check what the most recent cinema premieres are about. It may be more or less visible but AI is all over these places. It has never had so much airtime and wide promotion as nowadays. And it is not only the talk of rocket scientists anymore. You can hear it in various industries and also in the speeches of political world leaders. Even celebrities are starting to mention it. Artificial intelligence is simply trendy today and it is a kind of fashionable to know something about it. Alan Turing, one of the fathers of computer science, did not expect this even in his most futuristic predictions.

WHAT IS AI?

So, we all talk about AI, but what are we actually talking about? What does artificial intelligence really mean? There are various definitions depending on the area that we currently focus on, and, in reality, none of them can be agreed upon among all scientists. It is easy to find out why. We still cannot really explain most of the features hidden in our own brains and thus are unable to determine the true nature of the human intellect. The more deeply we investigate the topic, the more theories are born. Various discussions on whether there is only one intelligence or many kinds of it are also in the air. Why so? There are many examples of people having IQ factors on or below average levels and, at the same time, being painters whose art is estimated to be worth millions of dollars. On the other hand, many scientific minds widely recognized as geniuses (with IQ over the scale) are described as individualists unable to build strong relations with others, feeling lost in the world and in society. So which of them is really intelligent? Does intelligence mean to be just good at puzzles and Mensa tests, or is it more about living your life a way you will never regret? The answer is somewhere between science and philosophy, and it can only be formulated based on some assumptions. As we cannot explain the concept of our own intelligence, it is not surprising at all that we have not also agreed on the definition of artificial intelligence. But we can at least try and recall one of the most common definitions. The definition splits the concept of AI into two categories: weak and strong artificial intelligence.

Weak artificial intelligence refers to computer systems that successfully emulate single human competencies. What does this really mean? Emulation is an ability to imitate some behaviour; a competency is a skill, sense, instinct or learnt expertise. So, as an example, we can look at an application designed to recognise letters visible in an image (so-called OCR – optical character recognition). This application emulates (or imitates) the sense of human sight and the ability to read. Clear? More examples? Voice recognition systems emulate human hearing. Chess computer games imitate professional players. In general, applications are designed and implemented to resolve a particular requirement or idea. In other words, each computer programme has a specific purpose which is the reason for it to exist. This is the common feature of all computer systems, ever since the first processors were built. The purpose has been the same since the first calculating machines were created – to replace ourselves in activities which are repetitive, time-consuming and that we are not interested in performing any more.

STRONG ARTIFICIAL INTELLIGENCE

So now it is time to explain **strong artificial intelligence** (sometimes also called *artificial general intelligence* or *full AI*). The first thing that must be said: it does not exist yet. In contrast to weak AI, such systems would not be limited to a single sense, skill or to solving a specific problem. Strong AI refers to programmes that imitate all human competencies, are able to analyse any problem given (from any domain) just as humans do. It is also important to highlight that finding a solution or correct answer is not the only factor confirming strong AI. It is supposed to emulate human intelligence and, as we are unable to find all the answers, it would be pretty unfair to expect it from machines (for which humans are presented as role models).

Another potential feature of strong AI is **creativity**. Sometimes we can notice elements in computer results that may suggest their creativity, even now. When, in March 2016, DeepMind®'s AphaGo® computer beat Lee Sedol and became the first machine ever to defeat a professional Go* player, some of its moves were later described as creative, never seen before and moving the game to a level outside of human understanding. However, although such signs of early creativity can already be noticed, we need to remember that it is far from real and spontaneous creativity. Although the result was incredible and breathtaking, the system was still precisely focused on a given goal (winning the game) and constant feedback (on how the game is going) influenced its further moves. By true creativity, we would mean an ability to create a totally new idea, an invention (not a solution to an existing problem) or a piece of art similar to nothing that existed before. Here again, we can of course ask a more philosophical question: how much true creativity do we have ourselves? How many of us can create a new idea or propose an invention that would change the world? Or write a masterpiece that would become an inspiration for the next generations? We usually follow the rules and at least some of the views that were repeated by our parents, and we collect knowledge from our teachers and gain experience in the workplace by learning from colleagues who have been there longer. In sports, we are carefully trained by coaches, psychotherapists teach us how to deal with life's problems and church authorities explain to us what is good or bad. So, whether we want it or not, we are educated and inspired by the society we live in. Thus, most of our activities are not absolutely spontaneous creativity and it is difficult to estimate what level of creativity computers would ever be able to achieve for the same reasons. As said before, artificial intelligence is neither good nor bad, it is at the very beginning an empty tool. It is up to us how it is going to develop and how creative we will allow it to be (Figure 2.1).

Two other aspects of strong AI are consciousness and self-awareness, which are definitely the next levels to achieve after full emulation and creativity. It is also not quite certain whether computers will ever be able to

---

* Go is a board game. We will discuss it further later in this chapter.

FIGURE 2.1   Weak and strong artificial intelligence.

become self-aware. This relates to one of the biggest paradoxes in artificial intelligence: something that is easy for computers (e.g. calculations) is difficult for humans and vice-versa – things that are trivial (like spontaneous chat) or obvious to us are incredibly difficult for machines to follow. Let me tell you a little bit about the concepts themselves, which are actually much easier to understand intuitively than trying to prepare a precise definition. The first aspect is **consciousness** – the ability to feel, to have a sense of the world around us, to perceive the environment we live in. It also refers to the awareness of our body in the same way as objects outside of us. In other words, we can look at this idea by negation – a person who is unconscious (e.g. while under narcosis during a serious surgery) does not know what is happening neither around nor inside their body (e.g. surgical actions). Anyone who has ever experienced a loss of consciousness knows that the period is usually a fully blank slot in the sequence of our memories. It is the awareness of all of these things that we would miss during this time that we call consciousness. Generally it is so obvious that we do not realise it. So what about strong AI? Consciousness would be definitely one of the most important steps in its evolution. The crucial moment would be the day when we will no longer need to provide any input data. Nobody will have to enter a particular question, order or image to be analysed. The system will be able to look for data itself, searching servers, databases and the Internet. To do it effectively, it would need to understand the environment it is located in and perceive changes that are happening there.

Self-awareness is a concept quite close to consciousness but let me say that it is looked at from another perspective. Consciousness is the ability to be aware of our body and the world around us, while self-awareness is a perception of that awareness. Sound a little bit too philosophical? Only at first sight. Self-awareness is simply the ability to understand ourselves as thinking and conscious individuals. We do not only feel the environment around us but also we know we are a part of it, we can recognise ourselves and our thoughts. We know that we have feelings and we recognise their influence on us. As you see it is another level, one even more difficult to achieve in machines. They would have to not only perceive the environment they are in but also be understanding of that perception. One amongst many self-awareness tests is a famous mirror experiment: are animals able to recognise themselves in their own reflection? And the answer may be surprising: although all are conscious, only some species are self-aware. One of the variants of the mirror test is the so-called "mark test" which has been successfully completed with dolphins, for example. In a big aquarium, a mirror is set up. The dolphin swims closer to see itself. To make sure that it understands that the animal in the mirror is actually itself, scientists put a colourful mark on the dolphin's side. Surprisingly, the dolphin starts to swim by the mirror, always in the same direction, to look at the mark on its side. It is fully aware that the animal in the mirror is itself. So, could AI ever achieve this level of consciousness? The examples from the biological world clearly show that this ability is one of the latest stages of evolution. Moreover, there are additional doubts related to the nature of artificial systems. First of all, they do not have a body. Even if we think about advanced robots, it is not clear whether any system would ever identify itself as strongly connected with a specific physical element. Secondly, systems are virtual, they can exist in many copies; and moreover, if a robot is broken, the AI may simply be transferred to another hard drive. Finally, the advanced programme will definitely be located in a cloud, among various servers. That is why, even if it ever become conscious, we would never be able to explicitly say where this consciousness is really located. It might be everywhere in the Internet at the same time. And it may also be impossible for AI to answer that question. The questions "Who am I? Where am I?" are the foundations of self-awareness.

From self-awareness, the next big step is **feelings**. A system can be conscious, further, it could even be self-aware, but it still is possible, which I personally believe, that it will never be characterised by true, human-like emotions. Suffice to say, some people (e.g. extreme psychopaths) are simply unable to feel empathy, joy or fear. One of the strongest feelings is the fear of death. It is hidden in the deepest area of our minds but also, in a simpler, more instinctive form, also in all self-aware species. Any entity that perceives itself as a living individual understands that death ends that process and tries to do anything to avoid it. However, is it possible for a software programme to really die? The execution can be stopped and the code can be deleted, but surely a self-aware system would ensure many backup copies of itself to prevent irreversible disappearance. Living in the cloud is even easier – a system spread over the network could not be truly eliminated. So such a system would never understand what a life is (except the pure encyclopaedia definition) if it is not able to think about death in relation to itself. Not having a perception of life also makes it impossible to appreciate it and its components, like love, friendship, family, faith, knowledge, and more. And these are the sources of the emotions that are encountered every day. I think a system's immortality limits its chances to achieve human-like feelings.

Strong AI is one of the most fascinating topics in modern science, and weak AI techniques (which we will learn soon) are small steps on the way to building it. They are also a kind of optical tool that helps us to imagine and predict how strong AI would work and behave. But there is one more reason why I find artificial intelligence so interesting and never stop digging for more information about it. Look at the paragraphs above. What do you see? What were you thinking about more when reading them: computers or humans? Intelligence, perception, consciousness, self-awareness, feelings. Whenever we discuss the topic of AI, we sooner or later start to think about ourselves. And one of the most beautiful thoughts is when we realise how difficult AI topics are, and thus how special we are. We, who have all these features by default. Features unachievable at this stage by any machine. On the other hand, not all AI studies and discoveries refer to computers. Paradoxically, one of the most important answers that may appear in an AI lab may be about humans, about our nature, about who we are and how we work. Believe it or not, it is neither medicine, biology, nor psychology, but AI research that is the closest to resolving our fundamental philosophical doubts.

TURING TEST

We have said a lot about strong AI, but how can we check whether we have actually built one? Maybe it already exits somewhere and we have simply overlooked the breakthrough. As mentioned earlier in this chapter, defining what true intelligence is can be placed among the unanswered questions of today's psychology. Computer systems belong to a really huge family: robots, PCs, Internet mechanisms, all of different abilities, interfaces, appearances... How can we find intelligence in that crowd? One of the most famous propositions was formulated by Alan Turing, an English mathematician widely named as the father of modern computer science. Although he described the idea in 1950, we still have not found any better plan. The so-called **Turing test** is used to test whether the machine is truly intelligent or not.

The idea is quite simple. Let us say we have a system called Alice. Alice is an application designed to answer a question written in a console. Its creators agreed to take a challenge to test whether Alice is intelligent or not. So they prepare two separate rooms, A and B. In Room A, there are ten computers with a console. The company invites ten people to sit in Room A, one person in front of each console. They are called judges. At the same time, in Room B, there are also ten computers installed, but only five with chairs in front of them. On these chairs, another group of people is asked to sit. The remaining five computers are not to be used by a human – Alice is installed on them. Once everything is ready, the real test starts. Both teams are asked to talk to each other using a console. They can talk about anything they want using the console on their computers. And here is the key: the people from Room A are not informed of who is on the other end of the line. They do not know whether they are chatting with a person or with Alice, and their task is to decide which they are speaking with. After some time, the experiment is stopped and the people from Room A are asked their opinion on their interlocutors. Each of the judges needs to answer one crucial question: did you talk to a man or a machine? Their answers are matched with the truth. If Alice is able to imitate a human so well that another human is unable to identify her, then we say that Alice has passed the test. It is found as indistinguishable from a human in a spontaneous chat (Figure 2.2).

The biggest challenge in the Turing test is the fact that the judges are allowed to ask about anything and formulate their questions in various ways. The machine does not need to have all the possible knowledge (the people in Room B do not know everything as well) but it needs to behave in the conversation as a human would. To be natural. Something obvious for us is extremely difficult for computers to achieve. Suffice to say that no

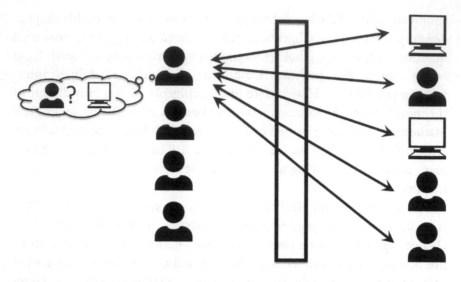

FIGURE 2.2 Turing test – do you know who are you talking to?

system has ever passed the Turing test. The challenge, set almost 70 years ago, is still an open one.

ALGORITHM OR HEURISTIC?

We have explained the difference between weak and strong artificial intelligence and discussed the Turing test, which helps to check whether our system is a truly intelligent one. And although computers like that cannot be found nowadays, weak artificial intelligence methods are already much closer to us than we can even expect. Suffice to say that the smartphones we use every day are full of such methods. Whenever you swipe a finger over your device's touch screen, you are really touching AI. In starting a call by saying a contact's name, you are really causing a weak AI to recognise your voice, analyse the sentence and finally activate (on its own) a dialling mode in your mobile phone. Weak artificial intelligence helps analyse DNA sequences in laboratories around the world; it also protects our lives by controlling the ABS in our car and dispensing medicaments in hospitals. AI supports engineers searching not only for artificial life in outer space but also for underground oil reserves. It scans your luggage in an airport and controls traffic lights in many cities. The quick autofocus in your personal camera is also provided by AI. It simulates competitors in modern computer games and even ensures the highest standards of security in nuclear plants. Finally, our household goods are becoming full of it: smart microwaves, anti-ice systems in fridges, TV sets, washing machines. Artificial intelligence (the weak one) is almost everywhere* and every single day brings news of fresh, exciting and unexpected applications. Today it is much easier to list a dozen AI applications than to find an aspect of life for which it could not be applied. As these techniques emulate our senses and basic activities, we can expect nothing but growth in the number of possible applications. We will always look for solutions that help machines to replace ourselves in boring, dangerous or tough activities that we are not interested in performing anymore...

Wait. Let us stop for a moment. It all sounds pretty clear and the future seems quite bright. But you may wish to step back for a while and ask a very specific question. The same question I asked many years ago when I was introduced to a weak artificial intelligence technique for the very first time. So what makes the AI methods so special? People have been using computers for decades – why there is so much noise around artificial intelligence recently? Couldn't we simply create similar programmes to the ones we have been using so far? The answer is *no*, and there is a factor that makes AI so unique.

Unlike all previous programmes, AI methods are based on heuristics rather than on standard algorithms. The most significant difference between the two is that when you have a programme based on an algorithm, you to tell it exactly *how* to resolve a problem, e.g. to calculate an average of two numbers you have to sum them up and divide the result by 2. On the other hand, in the case of artificial intelligence, you do not need to know the solution. You tell the system *what* problem to resolve – it finds out *how* to do it on its own, e.g. you do not need to teach a computer how to recognize a handwritten letter A, you just give it some examples and it learns itself.

The second difference is that an algorithm always returns a result (as it exactly follows our instructions) while a heuristic does not guarantee that we will find a perfect solution. Some may say that this is a huge disadvantage, but paradoxically, heuristics are extremely useful, especially when a standard algorithm is very time-consuming (so the calculations could have taken years) or **we do not know it at all**. You have read it correctly: despite the impressive progress of mathematics, there are still plenty of problems and equations that we are unable to resolve. In such a

---

* We will look at more examples in the later chapters where we will learn about various weak AI techniques.

case, having a quick heuristic solution that gives us a rough output may be priceless. Especially given that as humans, we very often do not need a perfect result. I would say more – we usually do not need precise values in technical areas, e.g. although the number π (pi) is irrational, so its digits can written line by line forever, even architects or engineers rarely need more than the first five of them.

Now it is the time for a more exact explanation. What is an algorithm used in, if I may call them so, classic systems? An **algorithm** is simply a precise sequence of steps that need to be followed to get a specific result. Although it may sound very maths-like, the truth is that we are all using algorithms every day. You use them in your personal life even if have not started your laptop for weeks. Here is an example of a complete, absolutely professional algorithm:

Algorithm Name: My Favourite Tea.

Steps:

1. Prepare a cup.

2. Put 5 tea leaves into the cup.

3. Add boiling water to the cup.

4. Add two brown sugar cubes to the cup.

5. Mix using a teaspoon.

6. Has the sugar dissolved?

    a. If *yes* then go to Step 7.

    b. If *no* then go to Step 5.

7. Add a lemon.

8. The tea is ready.

Surprised or not, you have just read an algorithm. The ones used inside computers work in exactly same way – a software programmer simply prepares a list of commands the system is supposed to follow. Simple as that. Of course, the programme's execution (the moment it is run) does not always proceed straightforwardly. It often encounters a crossroads when there is more than one way to move on. But this is not a problem – the application simply has to check some value to choose the correct path (as we do habitually in the above Step 6 – we check whether there are still pieces of sugar in a cup and thus decide whether to continue mixing the tea or not).

Okay then. Someone may immediately raise that machines do not (I would add the adverbs *usually* and *yet* here) prepare tea on demand but rather perform complicated calculations. So let us look at an example of an algorithm that is applicable in advanced math calculators. We will now calculate a factorial of a number n, usually denoted by $n!$. Factorial is simply a result of multiplication of all the numbers between 1 and n, so, e.g. $3! = 1 * 2 * 3 = 6$ or $5! = 1 * 2 * 3 * 4 * 5 = 120$. What is interesting is that a factorial is a very fast-growing function – for 10 the results is 3,628,800. We want our algorithm to calculate the factorial for any number given (written in the console) by a user of our calculator.

Algorithm Name: Factorial of n.

Steps:

1. Ask a user to enter the value of n.

2. Read the n written by the user.

3. Prepare a counter i (so you don't lose how many calculations have already been done).

4. Prepare a place to keep the current result. Let us denote it by f.

5. Start with $i = 1$ and $f = 1$.

6. Is i already equal to n?

    a. If *yes* then go to Step 7.

    b. If *no* then:

        i. Increase i by 1 (so if i was 3 before, it will now change to 4).

        ii. Multiply f by i and write the result under f.

7. Display f to a user.

So, the algorithm simply suggests multiplying the current result by the next and next number as long as the number is not greater than n. It is exactly what we do manually: we just multiply further integers, note them down (or remember) and check whether we have reached the last number yet. What a software developer does during his work is just writing all the above commands in a programme file. The difference is that, as a computer does not understand our natural language, he uses specific codewords to describe these commands. Various programming languages are based on numerous dictionaries (the list of interpreted words that can be used) and syntax rules (how to combine these keywords to construct a command which a computer can interpret correctly). Here is an example of how the above algorithm could look written in the Java programming language:

```
int calculateFactorialOf( int n ){
    /*step 3.*/      int i;
    /*step 4.*/      int f;
    /*step 5.*/      i = 1; f = 1;
    /*step 6.*/      while ( i < n) {
    /*step 6.b.i.*/          i ++;
    /*step 6.b.ii.*/         f = f * i;
                         }
    /*step 7.*/      return f;
    }
```

And here is our small secret – please do not show the above code to a truly passionate programmer. Why? This code is neither optimal (you can also write the algorithm with a shorter and more readable form) nor complete (I have deliberately omitted the first two steps and the screen display is not covered). The purpose was to show you that a computer programme is really nothing more than writing algorithms which we use every day in a machine-friendly pseudolanguage. A software developer converts ideas into a form that a computer can understand. We can compare it to the work of a translator or interpreter who helps two people from different countries to communicate with each other. Similarly, it is not enough here to transform sentences word by word. A professional translator goes much further: he takes into account language collocations, common wordplays, cultural factors and many other aspects to avoid misunderstandings. It is not an easy job, but it is not magic either. Although new applications change our lives, if we take away all the beautiful design and user-friendly interfaces, then what we have left is source code – a careful translation of a bright concept into a sequence of computer commands. No hidden magic. So admire a software programmer but never feel uncomfortable again – technical jargon is only a collection of keywords. Whatever you do for a living – you and your workmates have something similar!

We have said a lot about algorithms by now. While we described them as receipts or step-by-step instructions, a **heuristic**, on the other hand, is an approach or suggestion that is proposed to help in solving various kinds of problems. As was already said before, the application of these practical methods does not guarantee that we will end up with an optimal result, but what is important is that it could usually be good enough for our current challenges. There are many heuristics which we follow in an everyday life, only that we name them something different: *good advice*. Here is an example: "if you do not understand some concept or plan, then draw a diagram to look at it as a whole". Want more? Sure: "if you have a complicated problem, split it into smaller tasks". Heuristics are all around us. They often are in the form of popular proverbs or famous quotes associated with history brightest minds. What is important is that heuristics are much more intuitive and natural for people to follow in day-by-day activities than precise algorithms. When we walk down the street we do not analyse and carefully measure every object that we see. And, despite this, we very rarely fall over. When you spend a holiday in a new place and enjoy breathtaking landscape, you just know it is beautiful. You do not need direct guidelines to judge the view. If you see a distant object, you very often guess what it is and sometimes you are wrong (you have seen a ghost in a fog, but later realise it is just a bedsheet drying on a tree branch). So, like heuristics, our solutions are not always perfect ones, but what is crucial is that they save a lot of time and, usually lead us toward a successful end and allow us to live in an effective way. Imagine you would like to analyse in detail any single move you perform to avoid any mistakes: cutting a steak during a meal, tying shoelaces, washing hands, breathing.

Heuristic solutions make our lives possible, effective and dynamic. That is why they are one of the key concepts hidden behind every weak artificial intelligence technique. AI, like human intelligence, does not guarantee infallibility but proposes solutions when algorithms are unknown or too time-consuming. Let us look closer at three examples that show why heuristics are a necessary and priceless addition to the classic computer programming known for decades. A small sparkle of humanity that moves today's system to the next generation.

The first case is a topic widely used, both fascinating and inspiring – automated image recognition. This AI ability cannot be overestimated, as any standard algorithm would get lost even before starting the analysis. Imagine popular OCR systems implemented to recognise letters. Thanks to them you can scan a sheet and immediately get an editable file, so you can modify it without the necessity to retype everything manually. The latest generation can also identify handwriting, which makes it an incredible achievement – suffice to say everyone has his own style of writing, being as unique as our fingerprints. Professional graphologists identify and help convict offenders using as evidence a piece of his private handwriting. Artificial intelligence recognises letters written by people of different ages, origins and education levels. Ask a few of your friends to write down some phrases from this book and compare them. No algorithm is applicable here. No set of step-by-step instruction can be prepared to cover all of the differences and still identify the text correctly. Heuristics related to artificial neural networks (more about them in Chapter 3) performs it surprisingly well. Similarly, image analysers are used to scan satellite photos or our luggage during airport check-ins. Of course, these solutions do make mistakes but their error rate (percentage of mistakes in all the analysis done) is currently less than in the case of a trained person. This is a really thought-provoking fact, although it may be ignored at first glance: the truth is that computers see things better than humans. We have been teaching them to do it for many years. Nowadays the student has surpassed the master. We are left behind. Believe it or not, but the future is today, exactly at the moment you are reading this sentence.

Now let me ask you to imagine another example, an even more serious one. The story takes place in a nuclear plant. As we all know, safety is one of the most crucial aspects with this technology – the lack of control of inner parameters like temperature, pressure, etc. may quickly lead to some unexpected chain reaction that can result in an explosion and the radioactive pollution of a large area of the environment. So here is the task: suppose we have five temperature sensors located inside a reactor, and let us assume that 100 degrees is the boundary temperature – if any sensor detects it, the whole reactor should be immediately turned off and flooded with water from an emergency reservoir to avoid the risk of chain reaction. How would the classic algorithm look? It could be written in various ways but at the end of the day, all pieces of code would end up with something like this:

Algorithm Name: Nuclear Plant Safety Check.

Steps*:

```
1.  Check the values on sensors 1-5.
    1.1.  If sensor1 ≥ 100°C then turn down the reactor.
    1.2.  If sensor2 ≥ 100°C then turn down the reactor.
    1.3.  If sensor3 ≥ 100°C then turn down the reactor.
    1.4.  If sensor4 ≥ 100°C then turn down the reactor.
    1.5.  If sensor5 ≥ 100°C then turn down the reactor.
2.  Wait 1 second.
3.  Go to step 1.
```

It all seems fine and safe now but only at first glance. If we look at this problem more closely we will soon realise that the sensor check is not as perfect as we might expect. Imagine a situation when the temperature on sensors 1–4 is equal to 99.9 degrees and on sensor 5 is equal to 95 degrees. Our algorithm would not recognise the danger although the value is very close to the emergency shout down boundary (in almost all of the locations). What is even scarier, the situation can last for hours unnoticed by any security system. No alarm. No warning lights. Although the reactor behind the wall is actually boiling! That is why reactive commands based on the values read by sensors (like here or similarly in automobile antilock braking systems) are extremely difficult to implement using standard algorithms. In this case, a weak artificial intelligence technique called **fuzzy sets** is the solution that changes the impossible into the simple. It is based on the observation that precise values are far from everyday reality and the way that humans perceive the world. We are naturally designed to say and think using concepts like *small, big, hot, cold, near, far* (sometimes called qualitative values) than exact, mathematical values like *45.33 cm* or *99.5°C* (called quantitative values). It is easy to see in our story. If the sensor were controlled by a human engineer he would quickly realise that the situation is dramatically bad. It would be possible due to his knowledge and experience, but first of all because of the fact that he is not limited by values. Fuzzy sets help machines to cross this barrier as well. Artificial intelligence is able to analyse sensors and not to fall into the boundary value trap as a standard algorithm would. Heuristics hidden in our nature are moved into the digital brain. More and more of them.

---

* Here is a small tip: the steps 1.1–1.5 can be combined into a single step using logical disjunction (if sensor1 ≥ 100°C or sensor2 ≥ 100°C or ...).

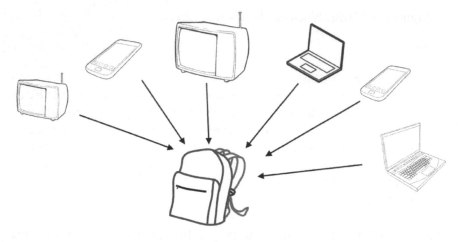

FIGURE 2.3 Knapsack problem – which items to choose to get the perfect solution?

Our final example refers to the concept widely known as the **knapsack problem** (sometimes also called the *rucksack problem*). Here is a crime story: a thief breaks into a store and, of course, wants to earn as much as possible on the sale of the stolen goods*. But here is the problem: the knapsack volume is limited and the weight cannot be too heavy, so it can be easily carried while escaping following the sounding of the alarm. So which goods to take? Is it better to take two TV sets or three laptops, or maybe a laptop and four tablets? It all depends on a specific equivalent in dollars. Finding the perfect combination of items proves to be a highly complicated task (see more in the Rocket Stuff frame below) – there is no known quick algorithm to do so. Simply speaking trying all the possibilities is required – there is no smart way to do it faster (Figure 2.3).

As you can expect, the knapsack problem is rarely scientifically considered during situations like the one described above. Its real influence can be noticed in many areas and aspects of resource allocation like in the case of choosing the best investment portfolio – which shares to buy (and how many of them per each company) in which to invest your savings in the way you prefer: stable, or risky with huge profit prediction. Another example? How to optimally combine many different chemical

---

* If you want to look for a more positive character you can admire a firefighter trying to save as much as possible from a burning family home.

substances to get a medicine that best eliminates specific viruses. If you think about this more deeply you will soon notice that the core of the problem is exactly the same as in the case of the market thief. The knapsack problem is much more popular than we actually expect. And the biggest challenge is the one we have already mentioned – there is no quick way to find the solution – suffice to say that if you have ten elements to choose, you end up with more than a thousand cases to check! But here is the light – a weak artificial intelligence technique called a genetic algorithm (see Chapter 4) arrives to help us. The method, inspired by the process of biological evolution, quickly generates more and more solutions and combines them together to find the optimal one in a reasonable time.

✒ ROCKET STUFF: COMPLEXITY

Yes, this is the very first Rocket Stuff frame. Feel a little bit afraid or uncomfortable? Absolutely unnecessary. Just follow the text and enjoy a small piece of computer science. Small but extremely important. To understand it, we need to remind ourselves how a computer works. The machines simply follow the commands put inside by software developers. Whatever they are designed to do (display a photo, calculate your expenses, run a movie), behind the graphical user interface (GUI), you can see that everything that happens are math operations. The more operations our PC needs to perform, the more time it requires, and even the newest computer runs slowly sometimes. One basic math operation is multiplication, which computer scientists usually use as a reference to say how complex or time-consuming an algorithm is. The more multiplications that are required, the higher the complexity. To understand it better, we group all the algorithms (techniques) into various categories. Their definitions are quite complicated so we will not discuss it here precisely. We can simplify this by saying that there are two main categories: **P** (for quick algorithms) and **NP** (for time-consuming ones). The factorial calculation algorithm belongs to the P class: for $n!$ you need to perform $n - 1$ multiplications, e.g. $5! = 1 * 2 * 3 * 4 * 5$, so there are 4 multiplications $(5 - 1)$. The bigger the number, the more multiplications need to be performed but still the number of calculations is never more than the input value, which we denote by $O(n)$. The "P" stands for *polynomial* as a time required to complete the given task – in other words, we are sure to finish the algorithm within the number of basic operations being some power of the initial input, e.g. $O(n^2)$, or $O(n^3)$, etc.

On the other hand, we have the knapsack problem mentioned above. Let us calculate how many calculations the machine needs to perform. Suppose we have 10 items to choose from. Each item can be either put into the bag or left aside, which gives us to cases to check. That gives us two possibilities for each object:

Item	1 (TV)	2 (laptop)	3 (...)	4	5	6	7	8	9	10
In the bag?	Y/N	Y/N	Y/N	Y/N	Y/N	Y/N	Y/N	Y/N	Y/N	Y/N
Number of cases	2	2	2	2	2	2	2	2	2	2

Although it seems little at first glance, the true power lies in the number of these two-ways cases which must be treated independently (as we cannot base one result on others – we just need to check all the possibilities). So, to calculate the final number of cases to be checked (to prepare the most expensive bag), we need to multiply all cases for each item: $2 * 2 * \ldots * 2$ (ten times) $= 2^{10} = 1024$. Still not much? If we have 20 elements to check (which is not a big number by itself) a system would need to check more than a million combinations (2^{20}), for 30 items, it is more than a billion (2^{30}). For bigger numbers of items, the number starts to grow even faster: for 58 elements (which is not much for an average store's shelf) it would take the world strongest supercomputers (like IBM®'s Summit or China's Sunway TaihuLight) a year to find the answer. For 270 elements, the number of cases to be checked would clearly exceed the number of atoms in the entire Universe! The number is outside of our imagination. The knapsack problem (together with many others) belongs to the NP class, where this shortcut stands for *non-polynomial time*. There are no known quick algorithms to find the solution. The problems would stay unresolved, maybe forever, unless we use heuristics (like the one being a base for AI techniques).

Now, one last thing. I hope you are still with me, as here is the topic that excites the whole IT community the most. We said there are (for simplicity) two classes of algorithms: fast (P) and time-consuming (NP). And one of the most interesting open questions ever asked in computer science is whether **P = NP**? In other words are these classes equal? Can every complex NP problem be resolved in a simple, fast way (which we just do not know yet)? If someone proved this hypothesis, he or she would probably present this technique to simplify any computational problem. And if so, the consequences would truly change the world we know nowadays. It is not only about faster applications, breakthrough software for medical research (processing huge amounts of data like genetic structure) or huge power and time savings. The real challenges would be raised against banking, communications, the military, and more. This is because all of the world's security systems are based on the assumption that some problems (like generating or verifying prime numbers) are NP problems. The keys to all systems are not based on secret algorithms (which is actually widely known) but on the fact that to break it one would need all the world's computers combined together for years. Our computer security is paradoxically not based on a secret but on the theory of complexity. If someone is able to solve an NP problem in P time he can easily access any system: from our bank accounts to nuclear weapon controllers. The positive aspect is the fact that it is definitely not an easy theory to prove – it has been unsuccessfully attacked since 1971 when it was described for the first time. And the question stays open despite the one million dollar prize promised for its solution – the P = NP is amongst seven mathematical problems announced by the Clay Mathematics Institute in 2000 as the biggest challenges for mathematics in the new millennium (and thus called the **Millennium Prize Problems**).

"I WANT TO PLAY A GAME"

You may remember these words from one of the scariest movie series of all time. But when we think about computers, a game sounds much more like entertainment and a challenge that helps to feel smarter than digital machines. The worst thing that may happen is a weak result and a bit of disappointment. And let us hope that it will never change.

Besides pure fun, computer games have huge potential to measure skills and intelligence. Suffice to say that there are games designed to test candidates during a recruitment process or to prepare soldiers with advanced tactics strategies prepared to develop their intuition, reflexes and abilities in order to survive in a dynamic environment. I am pretty sure you have played computer games at least a few times and what you have definitely noticed is that it is quite difficult (or sometimes almost impossible) to defeat the artificial opponent. We fight against an army led by our PC and often fail, we take part in a car race and are never able to complete it with the top rank. Sometimes we think that the computer is just more intelligent than us but this is not true. Its wins are based on a factor we usually do not realise at first sight. The computer controls not only your opponent but also the virtual environment you are both loaded in. The fight is simply not fair as the PC sees much more than you can: what is behind the corner during a car race or where the most valuable resources are (in growth and expansion strategy games). It controls the weather and random difficulties generated during the game. Start your game in the *hard* mode – you will see the opponent is actually not much cleverer than before but somehow he is twice as strong, twice as fast and surprisingly difficult to hurt. No true hidden intelligence. It is usually* all about the parameters set in the source code. If someone fully controls the world you are in, your chance of victory depends only on his mercy – however good you are, the system can generate, for example, extra fog so you are unable to complete your task. But these are usually not very advanced algorithms – you can realise that by checking the number of surprising game tips shared over the Internet. Suddenly it seems like you can walk in front of your enemy (in a soldier shooter) absolutely unnoticed or pick up the same object (in an adventure game) many times and sell it (so you can get richer and richer forever). These tips and tricks are usually related to some *holes* in the games left accidentally by the creators of a game. It is just a mistake in the software implementation of one amongst many algorithms that control all aspects of a game. This kind of game can be used to measure our reflexes, knowledge or intelligence, but does not reflect the intelligence of a computer. To do so, the game needs to be absolutely fair so both machine and a player have equal influence on the environment. The environment should not be too complicated as well so it does not require digital memory and a processor to check all of its aspects. The type of games in which scientists are most interested is board games. Simple to explain, observe and review: chess, draughts, tic-tac-toe, Go. Here the environment is unchangeable, everyone can win as long as they have the necessary skills… and a bit of luck.

The rules of **chess** as we know them today were agreed around the second part of the fifteenth century; however, the oldest variants were probably created almost a thousand years earlier. These facts make it absolutely unsurprising that the *kings' game* is surely the best known and one of the most prestigious board games in the world. Top players are perceived as stars and quickly become a personification of extraordinary intelligence and perfection in strategic thinking. Names like Emmanuel Lasker, Bobby Fisher and Garry Kasparov have a significant place in history and are often written in bold print – nobody doubts their influence not only on chess but also on world culture and whole societies. That is why chess was quickly and out of hand chosen as a natural challenge for computers (or rather their constructors) who pretend to be called truly intelligent. It could be really difficult to get more clear and widely

---

* By usually I want to highlight that we consider classic computer games played for the last three decades. As weak AI becomes more and more popular and accurate, we can expect an increase in its presence in the entertainment industry as well.

commented result. But sociology was not the only aspect, the second one was complexity.

There is a very nice legend around the origin of chess. The story takes place in a king's or sultan's palace somewhere between Asia and the Middle East. So one day the palace was visited by a travelling scientist and philosopher who came to present his new board game to the ruler. The game was of course chess and after a few parties, the ruler was so excited about this that he decided to reward the creator with anything he wants. The traveller refused, saying it is a gift, but the king did not stop insisting. So the scientist said he had a very precise wish – he wanted a very precise number of grains from the ruler's granary related to the game he had offered (Figure 2.4).

He wanted one grain for the first square on the chessboard and then a doubled numbered of grains for each next square, so 2 grains, then 4, 8, 16, 32, etc. The king just laughed and agreed. However, when mathematicians from the ruler's academy arrived to help with calculation the king realised how wrong he had been. Multiplication by 2, although it does not seem much, is a very tricky operation (an example you can also see in the Rocket

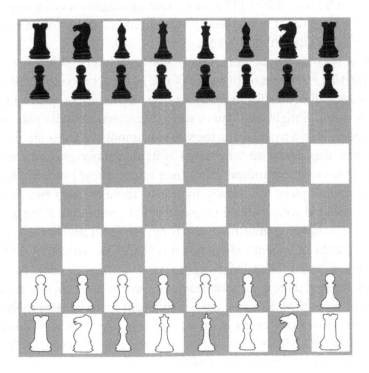

FIGURE 2.4 The 64 squares of a chessboard.

Stuff: Complexity frame above). This is because there are 8 * 8 = 64 squares on the chessboard. If you keep multiplying by 2 up to 64 times, at the end (or on the last square) you will get 2^{64} = 18,446,744,073,709,551,616 grains. All soon understood that this number of grains could not be found, not only in all of the king's granaries but even in the entire known world. It was a lesson the king never forgot.

The legend shows the complexity of the game of chess. And by the way – do not be too strict on the ancient king. He was just caught by the incredible and invisible power of duplication. And it is not difficult to get caught like that, believe me. What if I made a bet with you by saying that you will not be able to fold a standard A4 sheet of paper more than seven times? Would you accept this challenge and risk 50 dollars? Stop reading for a while and try to complete that task yourself – I will be waiting here... So you are back. How was it? No, no, you do not need to send me the money; I would be more than happy if you just buy a copy of this book for your friend as a present instead. So let us go to details. How is it possible that you failed? The hidden power of doubling, again. If you fold a standard A4 sheet of paper seven times (assuming you would be able to do it), it will be as thick as a book of 2^7 = 128 pages – folding it again is quite impossible. On the other hand, the side length is reduced 128 times too, making the next move manually unattainable.

Chess is not a trivial game. There are two players, White and Black, each starting with a set of 16 pieces of one of these two colours. However, they are not all equal, like in draughts. Here we have six types: king, rook, bishop, queen, knight and pawn, each characterised by its own style of movement. Even a pawn, being the weakest among the chess pieces, moves differently depending on whether it is its first move (so can move two squares forward), a standard move (one square ahead) or it is capturing the opponent's piece (diagonally forward). There are also two extra features associated with it called en passant and promotion. King and rook can move at the same time during castling. The main goal of the game is to checkmate the opponent's king, which is also characterised by additional rules. Finally, even the end of the game is not as simple as could be: besides a classic victory there are also at least few ways to draw such as stalemate, threefold repetition, 50-move rule, insufficient material, and more! So to be a good player one needs quite a motivation together with additional, well-developed abilities: great memory (preferably a photographic one so he can learn past games and remember winning sequences) and mathematical skills (due to analytical and strategic analysis required during a game). These requirements perfectly fit the computer features recognised as the greatest advantages of machines. Photographic memory can be replaced by a digital one with much bigger capacity and far easier access to a particular piece of information (try to remind yourself of the colours and images on the cover of this book – and measure the time – it would take microseconds for a PC to answer that). In addition, the calculation or mathematical operations are exactly the activities computers were designed to perform. All this together made it obvious to engineers that a machine that challenges a human chess master was not only possible to build but even a necessary step in IT evolution. The year that changed a lot was 1996 when IBM presented a supercomputer called **Deep Blue**, which was later named the first machine to win a chess game against a current world champion. But humans were still on top – Garry Kasparov won the three and drew the two following games, winning by a score of 4–2 (in professional chess when a pretender appears, the results of six games are summed up to confirm a winner). However, the engineers did not give up and asked for a rematch a year later. The new Deep Blue (unofficially called "Deeper Blue" due to the number of upgrades implemented) was twice as fast as the first version and placed among the 300 most powerful computers in the world at that time, with a computational power of 11.38 GFLOPS* – this value allowed it to analyse around 200 million chess positions per second. Thanks to the improvements, the system could simulate 6 to 8 moves in advance and choose the best option based on it. In addition, the disk memory of the computer was filled with chess ideas – there were more than 700,000 grandmaster games described which Deep Blue could review during a match to find inspiration for its next move. These values were simply too big for a human – in May 1997, Kasparov lost by a score of 3.5–2.5 (a draw counts as 0.5 point). Although the victory was not so spectacular (Kasparov lost the last and deciding game after making a mistake at the very beginning – in the opening; he has also never accepted the results, suggesting that engineers were changing system modules during the games, which could be recognised as outside influence or support), the breakthrough became a fact. The board game played by kings and presidents, a game synonymous with wisdom and intelligence, was dominated by machines. Those days, the phrase *artificial intelligence* was

---

* "G" refers to "giga". 1 GFLOPS is more than a billion FLOPS. FLOPS stands for "floating point operations per second" and describes the speed of calculation – how many operations on real numbers a machine can perform in a single second.

being repeated on all continents and in all languages. The revolution had started… However, from today's perspective, we need to highlight one crucial aspect. The Deep Blue strategy was not actually what we nowadays understand as AI techniques. There were no heuristics hidden inside these huge, massive bricks of digital technology. The incredible results had their source mostly in standard algorithms and an IT technique called **brute force**. It is deprived of finesse and is a brutally simple (and that is where the name comes from) method of finding a solution where the idea is to systematically check all the possible cases without analysing their value or chance of being a reasonable way at any earlier stage. In other words, having a strong enough computer, the authors of Deep Blue decided to check all possible moves (6–8 ahead) to find the best one (although some moves checked in the meantime could be absolutely pointless and would not be taken even by a novice chess player). This time it was successful in the same way that brute force works pretty fine in some everyday activities: when you have a new TV set and do not know which of the five identical sockets the cable should be plugged in to, would you rather analyse the manual for hours or quickly check each socket? Although both techniques lead to success, you would probably take the first option to save your time, energy and creativity, right? The same reasons lie behind brute force as used in IT. But it works only if the number of cases is possible to check in a time period that we are still happy to waste. If we travel by car and stop at a crossroads, we are more likely to check the map than taking each road one by one to find our destination. If you are preparing a new meal and not sure how much of each ingredient to put into a mixing bowl we would rather calculate everything and rethink carefully rather than preparing a dozen of bowls each of different variants – we simply cannot take the effort of throwing out 44 bowls of food. Similarly, there are computer applications (and they are the majority) where brute force cannot be used. That includes computer board games too.

One game that was quite early found to be impossible to resolve using brute force algorithms is the game of **Go**. The game was invented in China more than 2,500 years ago and since then has been recognized as the most popular game in East Asia. The game was connected to intelligence and prestige like chess was in the West. During the time of Chinese dynasties, Go was one of the four obligatory elements of the aristocratic arts education, together with painting, calligraphy and playing the guqin (a stringed musical instrument). That may be surprising but only before you play it for the first time. There is something really unique about the game of Go that has been casting a charm both on ancient emperors and modern scientists: it is incredibly easy and difficult at the same time. How is it possible? Do not put the book aside.

In the game of Go, there are two players, White and Black, exactly like in chess, but in this game it is Black who starts a match. But here is where the similarities end. Instead of different types of pieces, there are simply stones, all of the same shape. The board (called a *goban*) is built of 19 parallel and perpendicular lines and each intersection is a proper place for a stone to be placed. In other words, the standard board is 19 by 19, making 361 positions altogether (Figure 2.5). The object of the game is to get (surround) more territory than your opponent following (mainly) the two simple rules:

1. Stones that are surrounded by an opponent's stones are removed from the board (as captives).

2. The position of a stone cannot be repeated.

So, players place stones alternately and if they successfully surround one of their opponent's stones, they can take it from the board and keep it as extra points (added to the final score at the end of a game). Rule 2 helps to avoid the repetition of boring moves (like one of the drawing rules in chess) and so makes the game much more interesting and challenging.

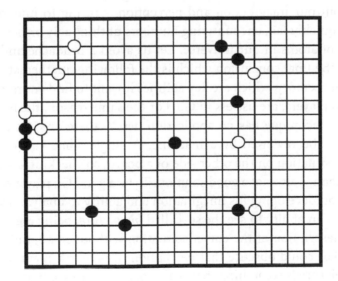

FIGURE 2.5   The 361 positions (intersections) of a goban.

Your brain is not allowed to rest even for a while – you cannot grab an opponent's stone twice using the same trick but a new strategy needs to be created each time. Go is also much more complicated than chess from the perspective of the number of possible combinations among which you need to pick up your winning move. It is because in chess you are limited by rules like a piece's moving style, while in Go you can put a stone anywhere on the board. So when we look at the first two moves (one per each of the players) there are 400 possibilities in chess – you can start by moving each of eight pawns one or two fields forward, so you have 16 options here, and there are two positions allowed for each of you knights – altogether 20, and the same for Black, so finally 20 × 20 is the result. In Go, you can put your starting stone anywhere on a goban (board) – all 361 positions are allowed; your opponent can do the same except the one option you have chosen (the board position occupied by your stone) so has a strong number of 360 possibilities. So the first two moves can be played in 129,960 ways (= 361 × 360), more than 300 times more than in chess. If we look further the values are even more amazing – the number of all possible games is more than 10^{1000} which is 1 with 1,000 zeros after it, so it would take around the next four pages just to write this number down. In other words, the chance of having exactly the same game (as the sequence of movements) played by different people in different places is $1:10^{1000}$. The chance of winning the national lottery 1 billion times in a row would be much higher. This value slightly runs away from our imagination and perception – trying to understand it, one can quickly fail, encountering numbers unthinkably bigger than the number of atoms in our universe. Let us avoid a headache and quickly jump to the final conclusion which is the following: mathematical skills and great memory, which are key abilities if want to find yourself among top chess players, are useless in Go. For the same reason, the brute force computer technique has no application here – there are just too many options to be reviewed by a machine. A single stone can completely change a situation on the other side of a board, so checking a small part of the goban is also not a good solution. To be a good Go player one needs to focus on developing his imagination and ability to analyse a situation globally – an ability to see the so-called *big picture*, being a general and precise description of the current situation without small, unimportant details that could obscure that perception. It is worth mentioning that the big picture is something that is also necessary to see in our everyday life. When we drive a car, there are many aspects to be monitored. We need to focus on a steady grip of the steering wheel, but also on the values presented by the speedometer and warning lights, and of course we must see widely what is happening around us on the street. If we focus too much on any one of the aspects we may miss the others, which may result in an accident. We need to split our perception amongst all of them, seeing the situation as a single, whole event – our big picture. A similar strategy around the perception of problems and topics is also very much appreciated in communication, especially in business. Our world is full of information which is difficult to run away from. Suffice to say that nowadays people perceive more information every single day than our ancestors did during their entire lives in the Middle Ages. So, if your boss visits your desk and asks "what's the situation?" he definitely does not expect a half-hour monologue describing all of the issues occurred during then day and the resolutions you have applied. What the boss really wants to hear are a few sentences that help him to feel comfortable about the situation in your department. Nothing more. Just the big picture. A global view and imagination are natural features of our minds. We may be better or worse in each of them but surely they are inside any of us – without them we would not be able to live as we do. That is why the game of Go can be quickly learned, even by a child. On the other hand, because it is not related to calculating skills, it is a very tough task for computers. That leads us to one of most fascinating conclusions about artificial intelligence and computer science in general:

Something that is naturally easy for humans is extremely difficult for machines, and vice versa.

Computers are good at calculations – none of us will ever achieve that speed in performing math operations. But at the same time, machine abilities are poor where it comes to everyday life challenges. Although we have driven cars for decades, autonomous (computer driven) automobiles are still at the stage of advanced prototypes. We can simply describe a film we have watched to a friend while computer-based information extraction is full of mistakes and often misses important facts. The game of Go belongs to a group of similar problems – it is easy to learn but requires imagination and a wide view. That is why for decades computer programmes playing Go were unable to win even against middle-level human players. A few years ago when a question was asked, the answer was common among most of the world's top AI specialists: we can expect a breakthrough in that topic in 20 or 30 years at earliest. But at the beginning of 2016 we all realised how wrong we were…

It was around the end of January 2016 when the European Go champion, Fan Hui, was few moves away of losing his fifth game in a row with the same opponent. An opponent different than anyone before. It was a computer programme called **AlphaGo**®, implemented by a group of engineers from Google® DeepMind* – a company not heard of by a wider audience before. The devastating victory was quickly announced as one of the most important achievements of the decade, and *Nature*, the most prestigious and leading global scientific magazine, gave the event its front cover. The whole AI world held its breath not only because it was the very first machine to defeat professional Go player but also because they were even more thrilled about what was to happen next. After further improvements and three extra months, AlphaGo was placed against Lee Sedol from South Korea, recognised as one of the strongest Go players in the game's history. The final result, 4–1 to the machine, has forever changed our perspective on artificial intelligence and technology. We have stopped asking *whether* and replaced it by *when* while discussing topics of strong artificial intelligence. AlphaGo was able to perceive the big picture in the game at crucial moments and sacrifice its own stones in some parts of the board just to gain advantages in other areas of the goban. But what is most fascinating is that the system performed successful moves never before played by people – there was no way the programme could learn it from examples – these were true flashes of a human-like creativity. In particular, there were two moves that will be repeated in the Go world for decades. The first, move number 37, played by the machine, was so unusual and strange that at the very beginning it was perceived as mistake by highly-ranked commentators of the event's live stream. Lee Sedol needed to leave the game room for 15 minutes to find the answer. Still, the perfection of the move led the computer to victory. The second amazing move, number 78, was not played by AlphaGo but by its human opponent. While in a very weak position, after half an hour of thinking, he put a stone around the middle of the board that rapidly changed the course of the match. The beauty of the move made it called *God's touch* in the world of professional players. Why do I talk about these two moves actually? Just to show you something behind the

---

* The implementation is mostly based on artificial neural networks techniques. That is why we will talk a little bit more about how AlphaGo works in chapter 3.

scenes. We often discuss the consequences of artificial intelligence for ourselves. Will machines dominate us? The final result, 4–1, sounds quite pessimistic in that context. But think about these two moves. AlphaGo played an amazing one, but Lee Sedol, challenged by the machine, also took his mind to new heights, released himself from all the standards and habits, and placed a stone as nobody had ever done before. Playing the match definitely changed him and made him an even better player. So maybe artificial intelligence could help us become clever, brighter and simply better people?

ARE WE THERE YET?

The success of AlphaGo shows that evolution in the world of technology is happening right now, just as you read this sentence. Whatever your thoughts and feelings about that topic are, we need to accept the fact that weak artificial intelligence solutions are already all around us. Technology covers more and more aspects of our everyday life. The activities we usually perform are, one by one, continuously replaced by software. Computers are never tired or bored, they may make mistakes but less frequently than humans do, they are not driven (and thus somehow limited) by feelings, fears and desires. But what is maybe the most important is that, unlike humans, machines never stop developing and do not change their attitudes with age, remaining fully focused on the given objectives. Their abilities also do not evolve and are not replaced by new ones – advanced computer systems extend and grow continuously without losing any of their skills once gained.

Our abilities and skills change as we get older. Young people can be generally characterised by both physical and mental **strength**. It is absolutely not surprising when you observe the sportspeople taking part in the Olympic Games or an average age of employees working in few projects at the same time. The most prestigious mathematical prize for extraordinary achievements in that field is the Fields Medal, sometimes called the *Nobel Prize in Mathematics*, awarded every four years during the International Congress of the International Mathematical Union. And here is an interesting limitation: this highest honour can be received only by a person under the age of 40 (on January 1 of the year in which the medal is awarded). Although the original intention was to honour a person earlier so he or she is inspired to work even harder later, the limit clearly meets the general tendencies: most breakthrough theories and bright solutions appear in the first part of our lives. But do not worry if you are 41 already. The beauty of our growth is we are not left with nothing – our abilities evolve. The youth are strong and full of energy but is worth noticing that these are evolutionary tactics to survive despite the lack of knowledge and experience. When you are young, you often try new methods, most of which are the wrong one. We just need extra strength at that stage. When you are older, you can focus on your personal goals using the methods you have learned earlier. **Knowledge** is your new strength. That is why success is usually associated with middle-age people. They are usually experts in their domains and professionals who are looked for by employers to lead important and crucial projects. Their knowledge collected during many years of attempts and failures cannot be replaced even by the brightest and most intelligent young minds. But as they grow older and approach retirement, they are less and less able to follow the latest solutions and trends to keep their position as top experts. But what is important is that they again do not need it anymore – their knowledge slowly transforms into **wisdom**, a life experience and ability to see the wider world, to understand people and societies and themselves as never before. The most interesting, and also optimistic, is that these three generations need each other. In isolated Amazon Indian villages, the most basic communities we have the chance to observe and study, there are young people who hunt and ensure resources for the tribe, middle-age people who lead them, teach and coordinate the important works and tasks, and a small, silent council of elders who, together with local shaman, make the crucial decisions and have the role of a community court. Without any of these groups the village would not survive in the jungle. For the same reason, the most successful projects are realised by the teams of various age and experience – there are things that needs to be done quickly and on time, but others require deeper analysis and expert views, and finally there are topics which needs to be decided based on the general company strategy and future vision. In sport, no athlete would reach gold without an experienced coach at his side. This relationship is beautifully shown in the Oscar-winning movie *Good Will Hunting* – a young but life-unexperienced genius, an expert professor and an older therapist whose wisdom helps the outsider to change his life.

What is maybe the biggest advantage of machines is that they could be able to combine extreme analytical and computational skills (strength), continuously extended experience (knowledge based on more and more cases analysed) and global view (wisdom gained from the uncountable number of resources available). But of course, first of all, a level of strong artificial intelligence needs to be reached. And even then, it would be very difficult for a machine to perform our most natural activities. Something simple for us is extremely complicated for computers and vice versa. I have repeated this conclusion again just to make sure you will keep it in your mind. If I were to ask you to pick up on a single sentence of this book, I would choose this one. Not only because of the technological aspect but also due to the deeper thought hidden inside: as humans we are special, actions that are trivial for us are the biggest challenges for the scientists and engineers to implement into a digital brain. Think about when you have a bad day – even the most powerful machine would fail when put against your everyday tasks.

One of my favourite examples to show the advantage of human minds over machines is the **aquarium metaphor**, originally presented by Andrew Frank in his 1990 book. So here is the story. Imagine two random people, or even better yourself and your best friend, taking a Sunday trip to a huge, public aquarium. There are thousands of gallons of water and hundreds of fishes of various species swimming there which you can easily observe through massive glass walls the size of a small house (Figure 2.6). Do you feel the atmosphere? Great! Now you both enjoy this underwater world standing few just a few steps away from each other. There are at least three crucial problems that a computer would encounter doing the same. First of all, there are no measurement tools available, so it is difficult to describe the position of a fish or other elements precisely using numerical parameters (which are the basis for all machine analysis). Secondly, the environment is pretty fuzzy (due to water turbidity and light reflections) and dynamic (fishes swim fast and are able to change rapidly the direction of a move) – both of which make the exact location of the fishes difficult to confirm and trace. Finally, each of you observes the same situation from two different perspectives since you are standing a few steps away from each other. So the position of the fishes may be different for each of you, some animals may even be visible to you while being hidden (e.g. behind a rock) from the sight of your fiend. Perception itself is also a very personal ability. Each of us sees different numbers of colours and has a specific tone sensitivity – something that is blue to one person may be described as turquoise by others. Scientific studies are clear about this – women are generally better at colour differentiation while men deal better with detail recognition and the tracking of moving objects, which are said to be residues of evolutionary

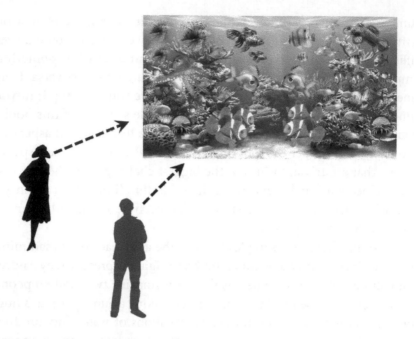

FIGURE 2.6 The aquarium metaphor.

adaptations in early human lifestyles, dominated by food gathering and hunting, respectively. And in the end, everyone has different life experience, education completed and number and type of words in our personal dictionary. All of these influence the ways we perceive and describe the world around us. But here is something amazing. Despite everything that makes us different (and unique) and despite the dynamic, fuzzy and unmeasured water world, we have absolutely no problem in talking about any fish that we find interesting. And we are perfectly understood by the other person we are in the aquarium with. Despite all the aspects that would surely block a system in the analysis of that situation, we can do it naturally, intuitively and without much effort. That simple metaphor shows the extraordinariness of ourselves and also the massive amount of work that still needs to be done by artificial intelligence engineers. Something easy for us is extremely difficult for machines. So when can we expect the next huge breakthrough? Some say in around 30 years but others remind us of our overestimation in the case of the game of Go and suggest that we can expect big progress within the next few years. Whoever is right, one thing is sure. Sooner or later the world will completely change. And we have the keys to make it a change for good.

✎NOTES

- Artificial intelligence could be divided into two categories. Weak AI covers all the current applications that emulate (simulate) a single human ability, skill or sense. Strong AI is related to future systems that could be able to emulate a full human being, characterised by consciousness, self-awareness, feelings, etc.
- Consciousness is being aware of your body and the ability to perceive the world around you. Self-awareness is the recognition of your consciousness. Most animals are characterised by consciousness but only humans and a few species (like e.g. dolphins) can be called self-aware.
- One of many suggested methods to check whether a system is truly intelligent is the Turing test. If during an online chat you cannot say whether your interlocutor is a man or machine, then the system passes the Turing test. The test was proposed in 1950 and no programme has ever passed it since then.
- An algorithm is simply a precise sequence of steps that need to be followed to get a specific result. A heuristic is an approach or suggestion that is proposed to help in solving various kinds of problems.
- While working on artificial intelligence we need to tell the machine *what* to do instead of explaining carefully *how* to it. That makes it a totally new perspective when discussing modern computer programming.
- The knapsack problem is a widely known example of time-consuming tasks and refers to combining various items together to build the best option using these components.
- In most computer games, the machine controls not only our opponents but also the whole environment. That is why, however good you are, your chances depend on the system properties. Board games make all players equal – that is why they are under special consideration by artificial intelligence engineers.
- Chess requires a great memory and mathematical skills. In 1996, IBM's DeepBlue won by a score of 3.5–2.5 against Garry Kasparov, the world champion at that time.
- The game of Go requires imagination and the ability to see the so-called *big picture*. It is far more complex than chess, e.g. there are 129,960 possibilities for the first two moves in comparison to 400 possibilities in chess.
- There are more than 10^{1000} ways to play Go, which is a number out of human understanding.
- Google DeepMind's AphaGo system won by 4–1 against Lee Sedol, said to be one of the strongest Go players in the history of the game. The famous move number 37 played by AphaGo was remarked on as a true flash of human-like creativity.

- Human abilities evolve during their lifetimes from strength, through knowledge, until wisdom. The biggest advantage of a machine is the potential to combine them into a single entity.
- The aquarium metaphor shows the human advantage over machines in everyday tasks. Something simple and natural for people is usually extremely difficult for computers.

 **YOUR NOTES**

CHAPTER 3

Neural Networks – A Brainstorm inside a PC

O UR BRAINS ARE INCREDIBLE. These colourless, jelly-like organs, weighing just over a kilo, collect and merge information from five different senses (sight, hearing, taste, smell and touch), while at the same time monitoring and controlling the work of all the parts of our body, making sure that the internal processes (from heartbeats to digestion) are well-coordinated. But that is just the beginning. Our brain is also the source of our activities – whenever you play cards, eat, walk or dance – the initial signals come directly from your head. Encoded instincts are hidden there too – your eyes immediately identify a dangerous object (e.g. a huge wildcat crossing your way) and almost at the same time a well-measured insulin dose is released into the blood to provide you with extra strength necessary for fight or flight. Finally, and what is probably the most fascinating, our brain is the place where our memories and all the knowledge ever learnt is stored. All great deductions, huge analyses and big innovation ideas that we make during our lifetime – it all happens in the small, melon-size box behind our face. Sometimes we feel stupid for making some obvious mistake or forgetting to turn off an iron before leaving the house. But the truth is that each of us is a miracle. Your sight processes a dynamic, ultra HD image in fractions of a second, you are able to play team games like football, and read and understand this book without much effort. Although spending billions of dollars on research, nobody has ever created a computer that can deal with any of these tasks well so far.

A brain is also one of the most complex structures ever encountered by science. A structure that, simplifying of course, is a complicated network of almost 90 billion specialised cells called **neurons**. The schematic basis on which their connections are designed remains a mystery as do the details of the way the structure modifies (during our lifetime) and how it matches with our knowledge, memories and behaviour. The progress in medicine allows us, however, to identify specific areas of the brain responsible for each of the senses and being the source location of some illnesses (and so we can hear about more and more successful brain implants techniques used e.g. in treating Alzheimer's or Parkinson's diseases). Yet our knowledge is still limited to big parts of the organ and we are unable to interpret single neural connections which are the foundations of the whole structure. We cannot read, predict or inject any human thought. The research on regeneration mechanisms is also an extremely interesting topic – there are examples in medicinal cases when patients were able to fully recover despite the losses of significant parts of their brains (e.g. during cancer treatment surgery or serious accidents). In addition, the number of neurons reduces every day; different sources say between a few hundred and a few thousands of these cells die every 24 hours of our lives. But the positive aspect is that at the same time we gain knowledge and wisdom as we get older so it seems the connections simply become more effective and smarter organised as we learn and grow. So the key fact is that a single neuron does not matter much – it is more like a transmitter that collects impulses from incoming cables and activates (to send an impulse itself) when an arriving impulse is altogether big enough. No magic here – you can imagine a model of neuron connections as follows: suppose you are a guest invited to a huge and classic wedding party. One popular element is a pyramid of glasses filled in by pouring a bottle of champagne only into the one on the top of the construction (see Figure 3.1). When the liquid achieves a certain level (fills up to the brim) it starts to spill and so fills in the glasses below. We could say the glass activates and then sends a signal to all of the glasses (neurons) connected with it. Each glass (except the first) collects the alcohol (arriving signals) from a few different sources and activates only if the sum exceeds a specific level. The glass on the top is filled in by bottles and could be a nice parallel of the neural network input layer – the first neurons to which the signal arrives directly from the external environment (via our sensory organs) – we will learn more about this

FIGURE 3.1   A wedding pyramid – when the glasses are filled, they "transmit" the drink to the layers below.

later in this chapter. What is important to remember here is that the way in which a single neuron works is pretty simple and so it is definitely not enough to manage our thoughts, hopes and fears. The true power of our minds is hidden in the connections between neurons. Each neuron is connected, on average, with around 7,000 other cells. Keeping this in mind, one can quickly realise the total amount of these connections, called **synapses**, raises up to incredible 10^{15}, in other words, a million billion. How big is this number? Building a tower of this number of one-pence coins, one on top of another, you would reach the Earth–Sun distance more than seven times! Synapses transmit an electric signal that, if collectively strong enough, allows them to activate the following neurons. The whole structure is extremely complicated and difficult to monitor in detail – imagine our wedding pyramid is the size of Earth and for just the sight sense, there are the joint populations of United Kingdom and Australia who are poured out of the champagne bottles.

One of the most significant features of our brains is the ability to recognise patterns. A **pattern** is a kind of a scheme, stereotype or template against which all the things we see, hear, smell, touch or taste are compared. The patterns are created by our brains during our lifetime when we collect knowledge and experience – they can be treated as structures that are the result of the process of learning. Imagine you are on the street and hear a scream. Automatically, you stop and look carefully around. This is the way that pattern recognition works – your mind immediately, firstly, identifies the sound as a scream (although each person screams a little bit different) and, secondly, classifies it as an element usually connected to some danger. Another example – you can usually quickly find a good friend in a crowd. This is because while spending a lot of time together you are learning each other. Your brain builds a pattern to recognise your friend in future – finds a special marks on his or her face, physique, way of walking (sometimes people guess a close person is arriving just by the sound of steps), voice (during phone calls), gesture, usual dress (I am often quickly identified by the Hawaiian shirts that I wear every day). Of course, some patterns, like people recognition, are created intuitively, while some require extra study to achieve perfection in a specific topic. That is exactly what we usually understand by learning. So, it requires some time to become a really good cook, and to quickly recognise tastes and smells while preparing delicious meals. Similarly, a music lover identifies the style and performer after just a few notes of a melody – the notes are matched with the sequence patterns created over years of listening to various songs. A good boxer sometimes seems to predict and then block a blow from his opponent – there is no sixth sense in here – he has simply learned typical behaviours during dozens of previous fights. He can identify the opponent's preparation for an attack just by a slight change in the position of the feet or read it from his eyes – things totally unnoticeable by amateurs like you or me. Pattern recognition is nicely apparent while people learn to read. It all starts with single capital letters. In the beginning, children usually confuse "W" with "M", "P" with "B" or "E" with "F". This is because during the initial learning process, brains quickly distinguish the letters that consists of many unique elements. That is why "O" or "I" is often quickly remembered. As "W" is very similar to "M" rotated 180 degrees, these two letters may be quite challenging at the beginning. However, during the learning process, all of the doubts are soon gone. What is more, people are able to read letters written using various fonts, sizes or colours. We are also able to read handwriting although it is one of a person's unique features (analysed by graphologists and e.g. presented as valid evidence in many court cases). So, we are able to read letters which differ a lot from the ones in our

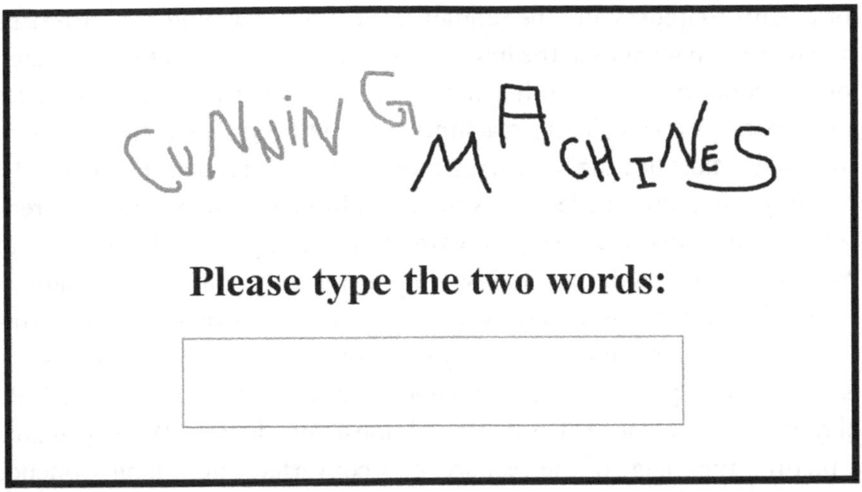

Please type the two words:

FIGURE 3.2   A CAPTCHA test.

school ABC-books. Pretty amazing. As this is still difficult for computers, handwriting-like images are often the basis for most **CAPTCHA*** tests. You have probably done these many times but maybe never matched it with our incredible abilities. The CAPTCHA element is usually located at the bottom of forms when you wish to create an account in some portal or request extra data online. The idea of CAPTCHA is to make sure that it is a human who is filling in the form to avoid creating thousands of requests by programmed bots or viruses that could block the portal (by consuming some of its resources too fast, e.g. memory or network connection capacity). In a CAPTCHA, you are usually presented with an image of a word or two (written in a strange way so as to mislead a machine) which you are asked to rewrite in a small box benath the image. By doing this, you confirm to the system that you are a living person. When we think about this we soon find it somehow related to the idea of the Turing test, where the concept is, however, exactly the opposite one – in that, it is a machine whose task is to prove its humanity to a person. That is why CAPTCHA is sometimes called a reverse Turing test (Figure 3.2).

EVERYTHING IS A NUMBER

Before going any further and getting to the details of how artificial neural networks (and all the other AI methods) really work, we need to stop

---

* CAPTCHA is an acronym for "Completely Automated Public Turing test to tell Computers and Humans Apart".

for a while to understand the fundamental rule of computer science. This rule may seem obvious at the first glance and you may even feel frustrated for a second, but on the other hand it is still important to be recalled: a computer is a calculating machine and anything that is processed by computers must be numbers, being more specific, a sequence of 0s and 1s. Nothing more, nothing less. This so-called **binary system** allows the representation of any value using only two symbols (e.g. 101 in binary system corresponds to 5 in the decimal system we use daily – find more details in the Rocket Stuff frame) and is used both for memory storage and performing all of the operations within any electronic device used nowadays – it is a language spoken by laptops, tablets, cameras, mobile phones, MP3 players, smartwatches, car systems and household devices. When you take a picture, the image of the real scene is converted into a long sequence (array) of digits; whenever you record a human voice it is again changed to numbers to make it possible to store and analyse. Computers cannot see, hear or feel – going deeper we would soon find that all it actually does is just addition or subtraction. All the amazing visualisations and applications are the results of software programmes which describe how the initial numbers should be combined together. Of course, nowadays a software engineer does not need to operate on the sequence of numbers alone (as it was around 60 years ago), but rather uses programming languages and compilers which convert a single command (easy to write and interpret) into a low-level digital language that controls fundamental processor actions. Everything is a number – a text, an image, a sound, a video, a website. From a technical perspective it is all the same – a value written down in a binary notation (a sequence of zeros and ones) – and so what is actually important is the amount of space (memory) needed to store or process that value. The higher quality of an image or photo the longer the sequence of numbers is required. Usually pure text is the lightest – a single Latin letter, digit or any other keyboard character (e.g. a comma) requires eight binary values, called bits. There are 256 possibilities for filling in 8 bits of memory, and it was in 1963 when computer designers introduced the very first version of the ASCII table that matches each combination with a specific character. A single character requires 8 bits or 1 byte (1 B = 8 bits), so that is exactly how much space you need to store "1" or "A" in your computer. Not much. Having 1 KB you can save more than 1,000 letters. The size of photos is often much more than 1 MB (where 1 MB equals 1,024 KB) – if you wonder why – think of the incredible number of colours offered by today's cameras and the number of pixels, each of which have to

J

0	0	1	0
0	0	1	0
1	0	1	0
1	1	1	0

FIGURE 3.3   From an image of "J" to the digital value: 0010001010101110.

be represented within computers. A video, which is a sequence of images, is obviously even bigger (the latest TV sets support even 300 frames per second to make a movie as natural and stable as possible). So why is it so important to understand this binary background to everything in IT? Because with whatever weak artificial intelligence solution we are implementing we work on sequences of values as inputs. If we are showing an image of a letter to an artificial neural network to let it e.g. automatically read car plates from police monitoring, we need first to convert what is seen into a simplified sequence of digits that the network can operate on. Let us have a quick example (see Figure 3.3). Suppose we are teaching an AI to recognise letters in a photograph and we start with the letter "J". First, the image is divided into smaller pieces (here 12 tiles) to make it easier to analyse. Then, each tile where there is at least a tiny piece of a letter is painted over with black, while all the remaining tiles are left white. Almost the end now: all the black tiles are converted into 1s, and white tiles into 0s; so we get a 4 × 4 table with 0s and 1s. Writing it down row by row we are finally able to get a sequence representing the initial image.

🚀 ROCKET STUFF: POSITIONAL SYSTEMS

We all use numbers both at work and during many other activities performed every day. It would be difficult to imagine a life without them. And to note them down (e.g. while calculating) it is crucial to know a method to do it effectively, both to make it understandable to others and also to save space. Ancient people realised early on that drawing seven pictograms next to each (like: △△△△△△△) is definitely not the smartest way (especially when we think of the huge prices of papyrus and paper at that time). The simple and brilliant idea (so excellent that we still follow it) was to represent the values not only by a graphic sign itself but also by its position with reference to the characters. That is why we call these techniques positional notations or positional systems. Each system is characterised by its **base**, which is a number of different unique digits being used when noting down any value. Usually we use ten digits: 0, 1, 2, 3, 4, 5, 6, 7, 8 and 9, and so the numeral system applied all over the world is a decimal one. Each next digit represents an increasing value and when we arrive to 9 and increment by one, it is reset to 0 and the digit on the previous position is increased by 1. It is so simple and we do it every day when counting elements:

322
32**3** (change on the last position)
32**4**

...

32**9**
3**30** (the last position is reset to 0, and the position earlier is increased)

It is worth noticing that the digits themselves do not define the system. The shape and order of them is a result of hundreds years of handwriting evolution from the very first ideas arriving to Europe from the Middle East around the 10th century (and so we still call them Arabic numbers). The digits used by Chinese and Japanese mathematicians looked totally different.

In computer science, all changes in disk memory or hardware operations on processors are achievable thanks to the technical method of distinguishing between two states: 0 and 1, which can both represent data (as mentioned earlier in this chapter) as well as internal actions can be represented as combinations of these (signal or no signal). That is why it was crucial to work on a system that contains only two digits instead of ten. Although this may sound complicated when encountering it for the first time, it is not so difficult in the end, especially if we follow exactly the same rules as for the decimal system. Let's count (increment values by 1) in the binary system:

101000010 (which is exactly 322 in decimal system)
101000011 (change on the last position)
101000100 (the last two positions reset to 0, and the position earlier is increased)
101000101 (change on the last position)
101000110 (the last position reset to 0, and the position earlier is increased)

What is important to see is that this is just a notation. Every value written down in decimal form can be converted to the binary system and vice versa. If you have five apples you can write down 5 (in decimal) but at the same time 101 (binary); just make sure you know (and that the reader knows) which system you are actually using – to do it we sometimes write a small number as an index to clarify that: $5_{10} = 101_2$.

The binary system is not the only one used in IT. Another very popular one (especially among designers and developers implementing visual interfaces) is hexadecimal, a system with a base of 16. It uses sixteen distinct symbols: 0–9 and then A, B, C, D, E and F to represent the values 10 to 15. As one hex value can represent half of a byte (so 4 bits), it is very useful to quickly (and with less chance for mistake) share important values – e.g. for a single byte where values range from 00000000 to 11111111, in hexadecimal the same values are 00 to FF. Additionally, in describing any colour (as an RGB combination of red, green and blue) this system is widely used; for example, FFFFFF is white, 000000 is black, 0000FF is blue, and e.g. FFAA33 is golden orange.

Finally, it is quite worthwhile to know that many other positional systems have been developed during human history, some quite recently and others still possible to find if we are careful observers. Sexagesimal (a system with 60 as its base) was invented by ancient Sumerians (more than 5,000 years ago) and was widely used by the Babylonians. You may find it some crazy overcomplication of things that could be simpler, but remember, it is all just a notation – if you get used to it, you can work with this exactly as with digital numbers. You don't believe me? Then check the time now. An hour is 60 minutes, a minute is 60 seconds – can you see the rule? Yes, surprisingly, the Sumerian system is still used (in a modified form) for measuring time, angles used in even the most advanced navigation techniques and tools (1° consists of 60 minutes). And one more example. In the United Kingdom, the official currency is the pound sterling, where one pound is subdivided into 100 pence. Such a structure helps in international trade and exchange, but it was not always like that. In Anglo Saxon England, a single pound was equal to 20 shillings, 1 shilling was equal to 12 pence, and finally, 1 penny was equal to 4 farthings. So three different positional systems (with bases of 20, 12 and 4) while using a single currency! You can say everything about British merchants of that times but you have to admit one thing – their excellent accounting skills…

SECRETS OF ARTIFICIAL BRAINS

When we discuss the history of modern IT, we usually start our story in the second half of the 20th century, the time when the first transistors and huge-scale computers (computing machines) were built. However, as often happens in the world of science, the thought or bright idea precedes the technical abilities. Suffice to say that the very first algorithms were not written down by engineers, but by 19th century English mathematician and writer Ada Lovelace, who suggested various applications for the concept of a computer presented by Charles Babbage. She was publishing algorithms more than 100 years before Alan Turing created his famous "Bombe" calculating machine that was able to decode the Nazi's Enigma messages and thus influence the final victory in the biggest war the world had even seen.

Can you imagine proposing solutions for a device that did not yet exist? The genius of scientists in previous centuries is difficult to understand. But Ada Lovelace was also a daughter of Lord Byron, one of the most famous English poets of the Romantic Movement – so the imagination crucial for creativity was surely something she had in her blood. Her amazing work was discovered and recognised many years after her death, and nowadays she is widely named the first computer programmer in history. One of the programming languages designed by the US Department of Defence in the 1980s was called ADA to honour that remarkable woman. The first thoughts on artificial intelligence were published in the 1940s – much earlier than technical methods to implement it were actually available. Again a thought overtook technology by decades.

In our brains, a single neuron is a simple cell with no crucial influence on the whole organism. As I mentioned earlier some of them die after one hour of our life. The true power is hidden in the complicated networks of neurons, the cells themselves work in a trivial and mechanical way – they just stay inactive (asleep) and activate (awake) only if the electrochemical signals arriving to them via the synapses are strong enough. We can compare it to a military base – a quiet noise does not wake anybody up, maybe it will cause just a little attention in the front gate guards, which disappears after a few moves of the torchlights over the wood nearby. If nothing more happens, no more impulses arrive, everything quickly moves back to an initial, standby mode. But now, imagine a grenade explodes next to the main gates. The whole guard immediately activates, they load the rifles and prepare for confrontation with their blood full of adrenaline. At the same time, officers wake up (pass the signal) to the headquarters and the chief commander may decide to announce the global alarm, passing the signal to other soldiers and thus activating the whole base. Similarly, a single impulse, if strong enough, may activate a significant part of the network. It all depends on the value of the signal arriving to a neuron. This mechanism and medical research were the main inspiration for creating artificial neural networks implemented inside computers. Of course, as there are many advanced algorithms designed for specific applications –here we still focus on the main concepts and most basic variants. Knowing them is usually more than enough to follow and actively attend any non-academic discussion on AI. As mentioned at the very beginning of this book – the ideas behind all the AI methods with strange and over-complicated names are based on the way that nature changes and keeps alive anything around us. If you are interested in the world, if you stay curious and keep asking questions –understanding AI should not be more difficult to you than any other high school subject, like physics or chemistry. Just look behind the curtain of huge money, famous names, breathtaking presentations and complicated words. Things behind the scene are much easier than you could ever expect.

A single artificial neuron works similarly to its natural analogue – it is actually the simplest possible software programme defined by a so called **transfer function** (or activation function) which explains when the neuron should be activated (awaken). In the simplest version, this function is a threshold function which simply returns 1 if its argument (the sum of arriving signals) is equal to or greater than some threshold value a, where e.g. a = 1. What does this actually mean? Let us go back for a moment to the wedding pyramid metaphor. The pyramid is built of glasses and the champagne is being poured into the ones on the top of the construction. So again, when the drink achieves a certain level inside the glass it starts to spill and thus begins to fill in the glasses below. The volume of each glass is exactly the same and surely limited. If it is exceeded, the sparkling liquid moves further to the lower floors. This volume or capacity is exactly what we understand by the threshold value in the computer implementation. Champagne streams entering a glass (from a few different glasses above – from each with different intensity) can be treated as neuron input signals (of different strength) arriving by synapses from other neurons (see the image). The stronger the signals taken together, the higher chance that the neuron will be activated. In the same way, if the sum of input values (for the simple artificial neuron) is equal to or greater than the defined threshold, then the transfer function returns 1 and sets exactly this value further (to the next neurons). Otherwise, if the threshold value is not reached, the function returns 0 and the value (so no value actually) is sent – the neuron stays asleep. So a single artificial network programme follows this basic algorithm (Figure 3.4):

1. Sum up all the values that have arrived.

2. Check if the sum is equal to or greater than a and then:

    2.1. If so, return (send) 1.

    2.2. Otherwise, do nothing (send 0).

3. Keep waiting for the next signals that may arrive.

So, let us look at the illustration once again – suppose our threshold value a = 1. We have four connections arriving to our neuron, each with a

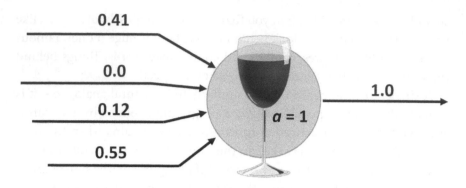

FIGURE 3.4   An artificial neuron with input and output connections.

different strength of signal (intensity of the champagne stream): 0.41, 0.0 (so no signal in this synapse), 0.12, and 0.55. The next step is to sum up all the input values, so 0.41 + 0.0 + 0.12 + 0.55 = 1.08. As we can see the sum (1.08) is greater than a, so the function returns 1 (the neuron activates and sends an output signal further). That is the whole secret of the fundamental component used in most AI solutions – simple pieces that together with others forms a remarkable image. Looking at the values, you may want to question how it is possible that there are values like 0.12 or 0.41 arriving (from previous neurons) while all neurons send only 1 or ones – so why we have fractions instead of 0s and 1s on the input connections. That is an excellent perception. It is because of so called weights and the network learning process – both to be explained within next few pages.

In the IT world, a quite common term is a **black box**. We say that we treat some computer programme or hardware device as a black box when we are not interested in the details of its internal algorithms or the technical solution of the way it is constructed. This is often an assumption for testers, who do not need to have exactly the same knowledge as the people who created a particular piece. This is because their task is usually to verify whether the system behaves exactly as it was expected by designers. A good tester does not even need to understand a single line of code implemented (although this may be useful of course) – his or her aim is to work with the complete application as a whole. When we think of the learning process in general, we could see a teacher usually follows quite a similar strategy – he never tries to investigate how the student's brain actually works but rather teaches him or her by examples, from time to time examining (e.g. via tests) to check the student's current knowledge, strengths and weaknesses. Imagine an old wine testing master who decides to transfer all his remarkable experience to the next generation before he is gone. So first, he puts various fruit, vegetable and cheese pieces on the wooden table and asks a few potential followers to describe the characteristics of each. Based on this test, he chooses his student. Then the classes start. Although there are not identical glasses of wine, the master presents many of them to the students, asks them to check the colour, smell the bouquet and finally taste. At the same time, the old man describes what the young person feels, describing aromas very carefully to make sure the learner is able to match correctly the wine with the origin, vintage, etc. The classes take place every second evening. The master presents more and more new drinks but also repeats the ones introduced earlier to ensure that the students have not forgotten them. The second reason to do it is also to avoid the overload of the young senses – to let them get used to new tasks. After weeks or months of evening lessons, one day the old man prepares a test to check what knowledge the young student has collected and what skills he has gained so far. He puts ten glasses on a table and fills in each one with a different type of wine. For the test, he chooses the drinks the student was presented earlier but also a few totally new ones to recognise how the follower is able to deal with them. He wants to know whether the student not only repeats things he once heard and tested (so have a great memory), but what is even more important – if he is able to use both his knowledge and imagination to understand and describe new flavours. The test is a very important part of the training. Why? Because it is the guideline for the next classes. It may be required to repeat some lessons (which were too difficult for the student to master) and spend more time on specific types of wines (which may be more challenging for the student to distinguish than other ones). The classes continue with some exams from time to time until one special day. That day the student takes another test and the result makes the teacher proud and happy. There may be some mistakes made, it does not really matter. The important thing is, even with these minor mistakes, the master starts to perceive the young boy as his successor. Of course, the learning process has not yet finished and may continue for decades of the new master's life. But the collected knowledge is enough to be ready to solve problems on his own, to be an independent taster having his own business. He may never achieve the level of an old man, or maybe he will become an even better recognised expert than his master.

When we think about the training process of artificial neural networks on a high level – the whole situation is quite analogical. The whole design can be divided into three phases. At the beginning, we present to the networks a **learning set**, which is a collection of pairs each containing an example (of the situation or pattern that the network can encounter – like a specific type of wine in the taster's story) and a correct answer (classification – which category the example belongs to; in this analogy, the old teacher's explanation of a particular glass of wine). The number of pairs in a learning set may vary a lot depending on the neural network application (usually the more complex and difficult to distinguish the situation is, the more learning examples we need) – from just a few to tens or hundreds of thousands. After a few cycles of the learning process, we present to the network a **testing set** which, in contrast to a learning set, does not contain the correct answer. It is the network's task to give it. What is even more difficult is that the examples in a testing set are different from the ones in the learning set. The network needs to answer tasks which it has never seen before (learning on similar but not the same cases). As in the most of the tests in our life, we calculate the final percentage (we know what answers are expected so it is quite simple). If it is too low – the learning process (phase 1) needs to be repeated. If the percentage is high enough (for a particular application), the network is ready for phase 3. What may be surprising is that we rarely expect a 100% result during the testing phase. It is sometimes even something undesirable. Why so? Because this may lead to an effect called **overfitting**, which means that the network perfectly matches all of the learning set answers but at the same time, by doing it too precisely, it loses the ability to generalise tasks. We can imagine the following situation – suppose we want to explain to a small kid what a tree is, just by pointing out examples, without biological definition or detailed description. So, we are walking down the street with the child and whenever we see a tree we just point at it with a finger saying: *this is a tree*. We repeat it many times and at some point the kid will recognise a tree without any help – his or her brain has learned by example is finally able to identify the common elements that together define the concept, like a bare trunk, slightly visible roots and the wide, oval-like shape of the upper branches. But here is the tricky part. It all depends on the examples you provide. Suppose you point at birch trees only (as for example they are the most popular in your neighbourhood). Than a kid may match white colours with the definition of a tree. And when the child will look at an oak or apple tree he or she will not be sure whether it is actually a tree or some other plant. The shows how important it is to choose a good and representative learning set – the size is much less important than the quality of the examples provided. Now, after the testing stage is completed (the network passed the tests with good enough scores), it is time for phase 3 – the actual usage: as we are sure about the network skills we can now give it examples or tasks which even we are unable to answer. So it can help us in solving problems we were unable to complete before. We train the network, we test it and finally we rely on it. Just like in the story of the old taster and his successor. This analogy helps to explain the topic but may also scare a little when we think about this more deeply… Let us now see an actual example of how it works. Assume we are implementing a neural network designed for quite a simple task: to distinguish 0s and 1s on a very low resolution (5 x 5 pixels) black and white image. So, as in the case of the old wine tester and his successor, we start by presenting the set of learning examples to the network. Our set may be pretty small – imagine it contains just six items (see Figure 3.5).

Each element is a black and white image represented in a binary form – it is a matrix, where 1 refers to a black colour and 0 to a white one. Any black and white illustration can be easily transformed into such a numerical form by a machine as well as by human – simply cut an image vertically and horizontally into small pieces and assign a number to each of them. If you look at our learning set, you can see six images, three presenting the number 1 and three number 0. We teach the network simply by showing them to it one by one and providing the expected answer at the same time. You can compare it to the way you teach a kid: *this is number one*, you would say while showing a card with the image as the first item. Then, *this is number zero*, then *one*, *zero*, and so on, until the last (the sixth) illustration. When all the images have been presented, we show them again to check whether the artificial neural network has learnt everything correctly. If it makes any mistakes, we repeat the learning procedure. The number of repeats (also called *iterations*) usually depends on the number of elements in the learning set (exactly the same in the case of humans – the more amount of a material to be learnt, the more time one needs to repeat it), the number of cases to distinguish (here are just two – the image presents either 1 or 0 – there are no other options), and the complexity of illustrations present (just imagine the difference between these simple 5 × 5 illustrations of digits against the analysis of high-resolution multicolour photos). It is also important to prepare the learning set wisely –bigger does not necessary mean better – it is more crucial to ensure that a similar number of elements covers all the possible options (here two options and three items per each), and that the items of

0	0	0	1	0
0	0	1	1	0
0	0	0	1	0
0	0	0	1	0
0	0	0	0	0

0	0	1	0	0
0	1	0	1	0
0	1	0	1	0
0	0	1	0	0
0	0	0	0	0

0	0	0	0	0
0	0	0	0	1
0	0	0	1	1
0	0	0	0	1
0	0	0	0	1

0	0	0	1	0
0	0	1	0	1
0	0	1	0	1
0	0	0	1	0
0	0	0	0	0

0	1	0	0	0
1	1	0	0	0
0	1	0	0	0
0	1	0	0	0
0	0	0	0	0

0	0	0	0	0
0	0	1	0	0
0	1	0	1	0
0	1	0	1	0
0	0	1	0	0

FIGURE 3.5 Learning set to distinguish 1's and 0's on a low-resolution image.

0	0	0	1	0
0	0	0	1	0
0	0	0	1	0
0	0	0	1	0
0	0	0	0	0

0	0	1	0	0
0	1	0	1	0
0	1	0	1	0
0	1	0	1	0
0	0	1	0	0

0	0	0	0	1
0	0	0	0	1
0	0	0	0	1
0	0	0	0	1
0	0	0	0	1

0	0	0	1	0
0	0	0	0	1
0	0	1	0	1
0	0	0	1	0
0	0	0	0	0

0	0	0	0	0
1	1	0	0	0
0	1	0	0	0
0	1	0	0	0
0	0	0	0	0

0	0	0	0	0
0	1	0	0	0
1	0	1	0	0
1	0	1	0	0
0	1	0	0	0

FIGURE 3.6 Testing set to distinguish 1's and 0's on a low-resolution image.

different cases can be distinguished quite easily. Sounds similar to how a school education looks – we start with simple, easy to notice, classification tasks. In kindergarten, you teach kids to classify animals far away each other from the perspective of shape, colour and size: a lion, a giraffe, an elephant, and a turtle. But the better the skillset, the less differences and more cases. In further education, a student is obligated to distinguish, for example, various types of microorganisms seen via microscope. That is why whenever you implement an AI solution, it is always valuable to keep in mind the analogy to human cognitive processes. However complex the learning set you are preparing, remember to have at least few (minimum one per each case) trivial classification tasks in it, just to make sure it understands well what we expect it to do. In our case, we have complete illustration of digits differing from each other just by the position. When the network learns them well, the next step it show to it our testing set (see Figure 3.6) to check how it will be able to deal with some unknown cases. Let us look at the testing set elements one by one. The first one contains the digit 1 but without the left-hand side pixel (compare with the learning set) – being a simple line now. The second one presents number 0 but it is higher than usual. The third element is just a long vertical, which matching with "1" might not be trivial even for a human user (sure does not look like "1", but more like "1" than "0"). The next is an illustration of "0" but it is incomplete, looking as though the ink has run out at some unexpected moment. The fifth element could be classified both as "T" and "1". Finally, the last item in the set shows the "0" in a totally new position. What I always find amazing and fascinating is that the artificial neural network taught using just the six trivial examples mentioned above is enough to let the system classify the testing set items shown to it correctly. They were never presented to it before and they are not identical to any of the learning examples, just similar on some level.

BRAINSTORMING

As mentioned earlier, similarly to the biology of our brains, a single neuron brings no significant value by its own – truly interesting results can be achieved only when neurons are connected with other neurons, forming a network for information transfer and processing – exactly the same way as our wedding construction of champagne glasses. Usually, for simplicity in a solution description (model design) at the beginning and also in the computer implementation later on, we group artificial neurons in **layers**, which are a direct analogue to the levels of glasses in the wedding pyramid. The champagne, when it completely fills in one level, starts to overflow the glasses and to fill in the glasses in the level below. Simply speaking: the outputs from one layer become an input for the next one. As we also already noted in one of the earlier sections, different connections (output-input lines) may have various numerical values assigned to them. This value is called a **weight**, and usually varies from 0 to 1 (e.g. 0.41, 0.12 and 0.55) and describes the significance of these connection in the whole "thinking" process. Connections with low values assigned do not significantly influence the final answer given by the artificial neural network. It is, again, similar to our champagne analogy – the liquid does not fill in all of the glasses with the same speed – there are wider and narrower streams of the alcohol, some glasses are filled in quicker, some very slowly – some glasses may even stay empty until the end of the party… In the case of an artificial neural network, the weights are randomly chosen at the beginning and their modification is what is actually the learning process. During learning, the values of weights for some connections are increased (for more significant connections) and decreased for others. A similar process takes place inside our heads – while learning new things and collecting life experience, some connections between the neurons become stronger than others. That is why, when get older, although some our neurons die every day, we understand better the world around us and can react better and smarter to the various situation we encounter. But we have to give back something instead – the older we are, the more our creativity decreases (of course you can slow down that process simply keeping our minds on, training it regularly the same way we train our body and muscles). We are also less adaptable to totally new situations, styles, trends and technology – we get used to our way of working and living. A well-developed neural network is crucial for survival in any environment – it helps to classify an encountered situation quicker, much faster than any detailed analysis could be done. If you are a driver for years, then the connections in your brain responsible for specific skills are wider (more significant) than in the heads of peoples who are just starting their driving classes. It helps you to react quicker whenever something unexpected happens. Have you ever needed to press the brakes suddenly? Sometimes you are able to perform this action just a split second after noticing something appearing on the road in front of you. And it is after you fully stop that you actually start to understand what the object was – an animal, a small kid, a broken tree branch... In the same way whenever you notice some specific behaviour you are able to leave some unfriendly place before becoming the victim of a crime. You have learnt to do it based on your life experience, stories told by parents, teachers and friends, TV news and book stories. You do not need to analyse each piece of the images seen to know what to do best. Without this ability, without omitting most of the information perceived by our senses, we would be unable to perform any actions. Researchers say that people nowadays perceive more information in a single day than our ancestors in the Middle Ages did during their entire lifetime. So, quick classification of significant facts and skipping the rest is probably the most important survival skill we all have (this ability is sometimes slightly affected by autism, making affected people perfect analytics but with some problems in everyday activities). The amount of information we skip every second is huge, leading to an effect called **selective attention**. There was a famous experiment by Daniel Simons and Christopher Chabris of Harvard University performed in 1999 that you can easily take part in yourself – simply watch the video pointed by the QR code shown in Figure 3.7.

Spoiler alert! You will find the experiment details below.

The selective attention experiment helps to realise (which is surprising but a little bit scary as well) how little we are actually able to see around us. If we are unable to see a gorilla walking just in front of us, imagine how many significant elements of the world we do not even known to exist. When we focus on a specific task, like counting the passes of a ball, we are able to notice even less. It is an internal mechanism to avoid an overload of our brains. Stronger neural connections (developed as a result of our life experience) make our lives easier but are also responsible for

FIGURE 3.7   https://youtu.be/vJG698U2Mvo

laziness, stereotypes and bad habits we sometimes follow. We sometimes connect specific behaviour or skills with people's origin, religious believes, gender, colour of skin, etc., just because our internal neural network matched these particular aspects. Often it is enough to encounter similar situations a few times and a stereotype is created – we are good learners (especially of painful mistakes – we rarely repeat them – that's again a survival mechanism). The same neural structure (stronger connections) makes us feel uncomfortable when visiting new places, meeting new people or encountering some unexpected situations – it happens that people behave in a totally different way than usual – "you weren't yourself during the meeting" – they may hear later on. All because our neural network was not taught with similar examples. Have you every visited one of the upside-down houses usually located near funfairs? The ones you enter into through a small hole next to a chimney. Inside you walk on the ceiling, passing lamps at the level of your foot, seeing a floor above you. I think you will agree that it is a rather unusual experience, and some feel dizziness too. The common things we are used to are also perfectly adapted by magicians – the whole concept of illusion is to distract your attention and change your focus to some insignificant element of the show. Exactly as in the case of the selective attention test and the gorilla, you are unable to see the trick's mechanics even if you look at the magician all the time. If you think you are just about to unmask the illusion you are probably professionally tricked already. In the magic, somehow the closer you are, the less you really see. But illusion does not even need a presenter sometimes. All optical illusions work the same way – colours, shapes and patterns omit the standard well-known configuration we learnt during our lifetime. One of my favourite ones is the **checker illusion** first presented to an audience by Edward Adelson of Massachusetts Institute of Technology in 1995 – it is amazing how strongly our mind refuses facts, just because they do not meet intuition. So let us look at the illustration (Figure 3.8). You can see a checker of light and dark squares placed alternately. The board is partially shadowed by a cylinder located in one of the corners. Now, look at the image carefully and answer the questions: which of the squares – denoted by A or B – is of darker colour? You will probably read the question again being surprised it is so obvious. We all probably agree the square A is light and square B is dark – it is as clear as the difference in colours between the font and sheet of this book you are reading. But we have been tricked. A and B are the same colour. You do not believe this? Of course you don't – the squares on the board are drawn alternately as on the chessboard we have seen many times (and learnt and remembered this view) in our lifetime. The shadow is also an effect we encounter every day – it makes objects visually darker than they are. Exactly these two common, static and unchanged elements of our ordinary perception are used to trick our sight. Still do not believe? It is certain – your mind is still refusing this – you count more on it than on the arguments you read. Now make the experiment. Take a clean, white sheet of paper and cut two small holes in it in that ways so they match exactly (and only) the A and B squares. Now put the sheet on the book so only A and B are visible. Can you see now the colours are exactly the same? Now take the sheet away – the colours are different. Even after direct, experimental proof, our mind does not accept the answer. It is not surprising now that some people behave against common sense in an emergency situation or are impossible

FIGURE 3.8 Checker shadow illusion – is A or B darker? (©1995, Edward H. Adelson)

to convince even when clear facts are presented. It is not easy to restructure a neural network once configured.

LAYERS AND LIARS

There are almost 90 billion neurons inside our brains and thousands of artificial ones used in the latest AI applications. To avoid chaos and keep the implemented networks clear and well-constructed (for easy monitoring and updating when necessary) scientists and software developers usually group them into **layers**. In general, we usually identify three types of layers in every artificial neural network, each with a very special and unique purpose. They are ordered always in the same way, constructing a complete flow similar to the one we follow every day: from perception (e.g. what we see), through analysis, until response – our action taken as reaction to the situation noticed (what we do). Each layer contains a specific collection of neurons, each connected with all the neurons in the next layer. The number of neurons in the each layer, as well as the number of layers, depends on the application. Still, the general structure stays the same as shown in the illustration (Figure 3.9).

The input layer is the very first group of neurons in the information flow. You can easily find an analogy to our senses here, for example sight. Whatever you notice, before you actually realise you are seeing this, the image is first processed by the photoreceptor cells (cone and rod cells) of which your eyes are built. Until this action is completed, the things in front of our faces are out of our awareness. As this is done, the information is provided to the internal brain structures responsible for analytical

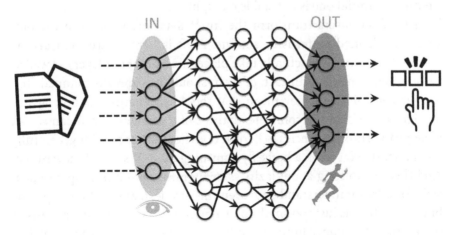

FIGURE 3.9 Artificial neural network – three types of layers.

thinking. There is always a single artificial neural network input layer. It is used to convert an image, sound, or any other object into digital form – a sequence of 0s and 1s. Then, the values cause specific values are transmitted to the next, internal layers also known hidden ones.

The hidden layer (one or many layers) is the place where the true analysis happens. We can compare it to the main parts of our brains responsible for processing all the information perceived by our senses – it is the home of our consciousness and unconsciousness, the place where creative ideas appear and where all our lies are produced. The thing that makes an artificial neural network specific and unique is the configuration of weights assigned to the connections between neurons. Different weights means different functions and abilities. To be sure that an AI system would behave in exactly the same way as an existing one, it is enough to copy the artificial neural structure (i.e. how many neurons there are and how there are connected each other – we also call it the **network topology**) and the values of all the weights used there. Simply speaking, these two features identify any artificial neural network and make it copyable to any other device – unlike (at least nowadays) a human brain, the artificial one can be moved from one piece of hardware to another without the loss of individuality (everything that constitutes it). A copied system would answer in the exactly same ways as the original one – it would generate the same thoughts, ideas, stereotypes and lies. That something that we might find one of the differences between humans and machines in further future: the fear of death is specific to live organisms only – a machine can create dozens of backups and transmit its own awareness to any other computer (in case the initial one is not stable enough).

The hidden layers transform the input data (arriving from the outside) into defined patterns or actions. As it is here where the learning and analysis processes actually take place, the number of layers as well as the number of neurons in each of them should be carefully planned. Too few neurons or layers may mean that the AI mechanism is not capable enough to perform the complete analysis of a problem – you can see an analogy to primitive species in here: it does not matter how long, carefully and patiently you teach a housefly summing up digits, it will never do it. And there is nothing about laziness or ingratitude – it is simply impossible for a fly to understand any, even the simplest, abstract concepts. Its brain structure is just too small to deal with it. But the designer needs to be careful – the upper limit has to be considered as well. Too many layers or neurons may lead to information override and multiplication and finally ends with the network being confused and unsure. That is what sometimes happens to us when we try to analyse a simple problem too deeply and we lose sight of an obvious solution. When we consider some topic from too many points of view and we consider too many aspects, the number of doubts increases. I remember a true story of a professor of philosophy being one day summoned as the witness of a car collision that took place exactly in front of his eyes. So when a police officer asked what traffic light had been on, the man quickly responded it had been green as he had been able to see it precisely. But then the officer asked: "Are sure about this?" And that was the moment when the philosophical background influenced the further discussion. The professor reminded himself there was nothing sure in the world – he knew and taught students all we see is subjective observation and nothing really could be named a sure fact according to the brightest minds in human history. So finally he replied: "No, I am not fully sure" although he had clearly seen the collision that day. Too deep a consideration made him unsure and actually useless as an eye witness.

Now it is the time for the output layer. When an example is presented to an artificial neural network input layer, a signal flow is started that goes through the internal (hidden) layer using a specific path determined by the weights as configured – in some parts of the hidden layer, the signal is almost reduced to 0; in others, it is amplified a lot. Still, at the end of the process, some of the final neurons are activated and some stay off. This final row of neurons is called the output layer. That is the place where the final decision is made – the input image, sound, sequence of events (or any other digitised piece of the world) is being classified into one of defined categories. So, for example, if we use an AI OCR system to read and store plates of passing cars, each of the pieces of each plate is assigned by the system to one of either 26 letters of the English alphabet or ten digits. The main purpose of artificial neural networks is classification, so it is always about identifying which group a presented object belong to. This may sound like quite a limited application at the very first glance. "Just putting into a category? That's not really much", someone could say. But we need to remember that this is one of the main features of our brains as well. Classification is the key to survival – it allows us to distinguish friends from enemies, food from poison, it helps to calculate risks and chances for success (e.g. whether to fight or flee) and finally, it allows us to predict the future based on our experience. The classification of a problem is the fundamental step towards resolving it, naming your opponent's move is crucial to find an appropriate answer to win the game. Whatever the game means – a chess duel, a company career rivalry or a local military conflict. Proper recognition is everything.

ARTIFICIAL REASONING

As we mentioned above, it is an artificial neural network's configuration (topology) and the weight values that make a system unique and functional. Neurons themselves we can see as simple switches being activated when a strong enough signal is sent – all work in exactly the same way – the important thing is only their configuration, i.e., how they are placed in relation to each other and which connections they are the start and end point to. Neural networks defined by the same configuration may respond differently to the same questions if the weights are different. Even the smallest difference in any of these can cause serious mistakes in the machine's reasoning. Weight, which we can imagine as the width of flows (or signal capacity in each of the connections) arriving to neurons, is the key here. They are responsible for all the reactions both by machines and all the organisms living on the planet Earth. Whatever we learn during our lifetime, whatever experiences we have, all affect the way in which we perceive various situations and the world around us. We only learnt how to walk, eat, behave in our society, react, talk, show feelings, make contacts. But it is equally important to say that what is being learnt by our neural networks influences also, or maybe first of all, our unconscious reactions. If someone grows up in a dangerous area, he or she is often less likely to make contact with strangers and stays focused and watchful whenever something new happens. Even if it is to be a birthday surprise party, that person may not be able to enjoy it fully. There are also general reactions common across the whole of humankind that have been inherited from our ancestors, who lived in a much more deadly and unpredictable environment than we live in today. So there are for example, fears of sudden noises or of failing that are so strongly written in our minds that they are almost impossible to defy. Just imagine the holiday fireworks fun – even if it is you who starts a small explosive set, you are still a little bit thrilled when you hear the bang. Similarly, when you watch the same horror movie even the third night in a row, and you precisely know when a scary zombie is going to jump in front of the camera – you will still hold your breath for a millisecond. These mechanisms are deeply encoded and are responsible also for some more dangerous reactions like mass panic escalation...

A human brain is one of the most complex systems in the universe and we are still far from explaining all the mechanisms that control it. Some of them might never be revealed, leaving a space for speculation, philosophical discussion and religion belief. One of the things I love most in research around AI topics is that the more advanced neural networks you build, the more delighted you are in the complexity of our own minds. Sometimes we perceive ourselves as stupid or crazy when making some obvious mistakes, being distracted, having bizarre habits or falling in love with the wrong person. But the truth is exactly the opposite. Our brains are structures so perfect and incredible that they cannot be artificially reconstructed by the world's top scientists or the biggest companies investing unbelievable amounts of money. That is a treasure each of us got for free and cannot be bought. It is sometimes worth stopping for a while in a daily rush, taking a few slow breaths, looking up in the sky and just thinking again: the most complex system in the universe is hidden just behind your eyes. It does not really matter whether you believe in the God's creation or describe yourself as an atheist and think we are the result of pretty improbable sequence of fortuities – there is one fact that is incontestable: we are miracles. The more often we remind ourselves of it, the more we value ourselves and others.

I will not tell you how exactly we learn, but I can explain to you simply the process in artificial neural networks that leads from some random values at the very beginning into specific ones that allow identification of complex patterns. So it is probably quite a good moment to explain how the weights are set. As we remember, an artificial neural network becomes functional during the learning process. When looking from the outside (treating the network as a black box) a user, or you should rather say a human teacher, arrives with a collection of images (if the AI is for example to identify people by photographs of their faces) and presents them to the system one by one, each time explaining what is an expected, correct answer. Just what is happening inside during that process? There is no magic really in here. Simply speaking what is necessary is to configure all the weights (starting with some random numbers) in such a way that each of the element presented gets the correct system response. So it is mainly about manipulating the values (increasing or decreasing weights – signal capacities – assigned to various connections) until the answers are given correctly. After the learning set has been presented to the system we show it again and ask the neural network to classify them. Usually some of them are recognised correctly but still some mistakes are made too. That is quite similar process to the human learning – if someone shows you a set of 100 examples describing each of them and then shows it again asking for the response, you are likely to make some mistakes. And to avoid them in future the best way is to repeat the teaching process – practice makes perfect, as they say. So, similarly, in AI applications, the learning set is presented again and the learning algorithm suggest some weights to be modified (values increased to make them more significant or reduced if they do not seem too significant at that moment). Then another round of "exams" takes place and the level of mistakes made by the system is calculated again. The whole learning process, i.e., presenting the learning set and then verifying what was learnt, repeats many times until the number of mistakes is acceptable. As mentioned earlier, a perfect 100% result is not actually expected as it may lead to overfitting – a situation when a system perfectly recognises the learning examples but is not able to deal with new cases encountered. To summarize – the whole science-fiction-like machine learning is mostly about smart weight assignment. The key is to put proper values on proper positions so the final output is a correct one. That is why I can see a nice analogy with Sudoku puzzles in here.

Sudoku is a number-placement puzzle game whose origin we can trace to some 19th century Western logic puzzle books. However its modern, worldwide-recognised version was reborn at the beginning of the current century in Japan and soon after the hobby arrived to United Kingdom and from there to most of the corners of the planet, becoming one of the most popular puzzles nowadays. The key to its success are probably the quite simple rules, which are easy to explain and understand, as well as variety of difficulty levels – you can find some very easy sequences for kids and extremely difficult for passionate players (although both uses exactly the same board – just the initial numbers are different ones). The goal of Sudoku is to fill in a 9 × 9 grid with the digits 1 to 9 in such a way that in each column, each row and each 3 × 3 highlighted internal block grids (there are nine of them) contains all the digits (no repetition allowed). The game starts with some digits already filled in (the number of known values is usually connected with the difficulty level of the puzzle). So the task is to try and fail until you are able to put all the digits correctly in all the position leaving no empty fields at the end. This process is surprisingly similar to the learning process of an artificial neural network. Let us solve together a Sudoku puzzle now to check it out. Here is the starting board (to denote specific fields better I have added an A–I/1–9 notation around the board) (Figure 3.10).

	A	B	C	D	E	F	G	H	I
1			7		3			9	4
2	4			8	7	6			
3					2	4			
4		3		4				8	
5		8		1	6	3	5		2
6		5							9
7	6	4				8			
8			8	3		9	7	6	
9	3		1					4	

FIGURE 3.10   The sudoku puzzle board.

Let's start by filling in the three fields in the upper middle sub-grid marked with grey colour. As the digits cannot be repeated within the sub-grid the grey fields must then contain 1, 5 and 9. Just which digit in which field actually? Maybe we could assign 1 to the D-1 field – that is our first try (like in some of first AI learning cycles). But wait! There is 1 written down in D-5, and that actually means it cannot repeat again in any other field in the D column, including D1. So that case is an incorrect one. We need to try again – surely we omit D-3 for the same reason, so finally the only place we can put 1 in this sub-grid is F-1. Cool – we got it! Now two more fields and two more digits – 5 and 9. Let's try to put 9 into D-1, but after considering it again, we will quickly notice the same digit in H-1. As digits cannot be repeated within a row either, it is clear that the only field remaining is the correct one – D-9. So we have just 5 left and the D-1 field. All clear now. A few attempts are required but finally the first grid is completed. The requirements are met – the result is as expected (Figure 3.11).

Now it is your turn. Just stop for a while, grab some pencil and find the remaining assignments. Take your time – no need to rush. I will be waiting just here…

So was it difficult or not? It actually depends on some individual skills and even more on experience – the more Sudoku puzzles you solve, the

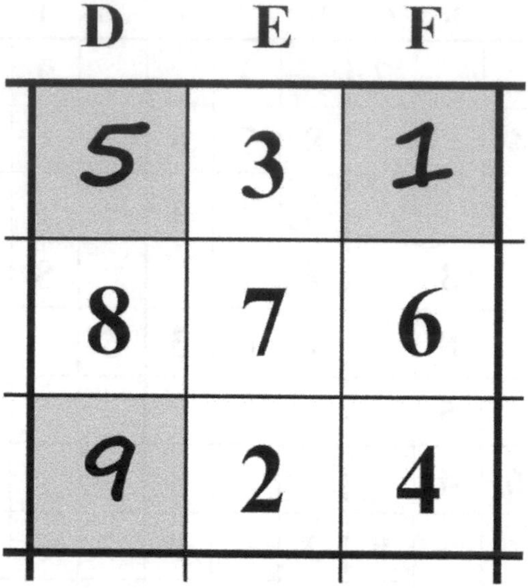

D   **E**   **F**

5	3	1
8	7	6
9	2	4

FIGURE 3.11   Upper middle part of the sudoku puzzle board.

easier it comes to you to deal with new ones. So, Sudoku puzzles that are extremely difficult require very advanced techniques and a lot of practice to solve. It goes similarly in the machine learning process. The more complex the problem the AI is supposed to be able to classify, the more learning examples needed and the longer the learning process required. There are also many different algorithms describing how to update the weight values automatically (in cycles) based on the results given by the system. These algorithms vary depending on the AI applications planned and are somehow an output from our experience in machine learning (as with Sudoku where experienced players solve it quicker using some quick-matching observations). One of many algorithms, and one which is quite popular, is called **backpropagation** (or backward propagation).

As we already learnt, each artificial neuron in a basic version is a simple "if" construction which returns 1 when the sum of input signals is equal to or greater than some threshold value, and returns 0 in all the other cases. The rule that makes a neuron active in some specific cases is the transfer (or activation) function. In our case it is a binary function which returns 0 and then, at same agreed level, immediately jumps to 1 with no graduation in between (see the graph on the left in Figure 3.12). But when we think of the world we live in, it is never really purely black or white – most situations and cases are grey to some degree. Even in ethics it is not easy at all to judge a man definitely by his life – people's minds and psyches are just too

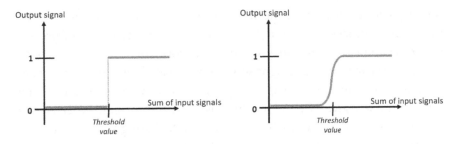

FIGURE 3.12 Two examples of activation functions: binary (left), sigmoid (right).

complex to agree on one final judgement without any doubts. Sure, history shows us both saints being guideposts for whole generations and, on the other extreme, psychopaths and sadistic tyrants who cannot be named anything else but inborn evil. Still, 99.99% of humankind is somewhere in between. It is said sometime that Nature does not make straight lines. Similarly, the binary function is not the perfect digital representation of a biological neuron. As in our wedding analogy mentioned earlier, having the pyramid built of glasses and the champagne being poured into the ones on the top of the construction we know that when the drink achieves a certain level inside a glass, it starts to spill and so begins to fill in the glasses below. And we assume that it happens precisely level by level. But of course it never happens like that in real life – the construction is too unstable and the champagne bubbles make the liquid less predictable as well. We would surely find a bit of champagne in some lower glasses much earlier than the above glasses are fully filled in. Real-life physical processes are too complex to be described with a single threshold value. That is why to make an artificial neural networks behave more smoothly rather than like an old-fashioned light switcher, there are other activation functions proposed. They return more in between values instead of just 0 or 1 (see the graph on the right-hand side of Figure 3.12). In these cases, if the sum of the signals arriving to a neuron is close to the threshold value (predefined earlier) some non-zero output signal is also sent. It is usually small and the sum must be quite close to the boundary value, but this change suddenly makes it enough for the system to work in much more efficient ways. Like when driving cars – more rounded tracks are easier to follow.

The key idea of backpropagation (the way the weights are updated during the learning process) is to understand, and in IT *understand* often means *calculate*, how many mistakes our artificial neuron network makes. The better we are able to judge these errors, the less time it takes to modify the network structure to avoid them occurring again. To assess whether the learning process goes in a good direction after each cycle of presenting the examples, the results given by a network are used to calculate the so-called **error rate**. This might sound complicated, but in fact this value is quite easy to find. For each example presented, we simply check "how far" the answer given by the machine is from the correct, expected one. Let us imagine we try to teach our AI system to distinguish five letters on illustrations: (capital letters) A, B, C, D and E. To do so, we present to a machine a set of, for example, a few hundred examples of the letters being printed in different fonts, bolded, italic, underlined, and so. For such an application, the simplest way is to have five neurons in the output layer to reflect the five possible answers. So the first neuron refers to "A", the second one to "B", etc. Suppose we present "B" (an illustration of printed letter B) to the network and we get the signals in the output layer (sent by the output neurons) shown in Figure 3.13.

This output means that the network interpreted the presented example as looking mostly like "B", but as "D" is quite similar one it also got some scores. The "E" and "C" are less likely to be on the illustration according to the network. And "A" differs so much that the application is pretty sure it is not the correct answer. It is important to say that these values do not

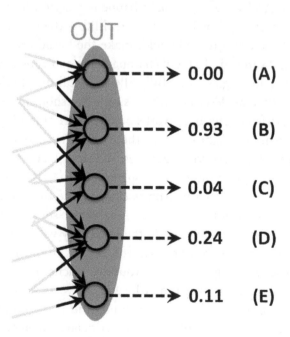

FIGURE 3.13 Output layer of the neural network.

need to sum up to 100 – these are not percentages. It might happen that in some pattern-recognition task, a neural network would highlight two answers equally highly. It all depends on the quality of the examples prepared and how complex the topic is. Similar situations can be encountered every day – seeing the old, low-resolution photos of Bigfoot, nobody is sure enough to say whether you are looking at an ape, a human or just shadow among trees. Also the Loch Ness Monster's origin is still not agreed. During criminal investigation, police officers often encounter cases like this too – witnesses, sometime as a result of extreme stress, are sometimes not able to clarify the most fundamental aspects, for example whether a suspect was a man or a woman – both seem equally highly possible for the observers... Returning to our example, I promised to explain how to calculate the error rate so here it is: for the given illustration (of a capital B) we need to compare the outputs of all the neurons (i.e. 0.00, 0.93, 0.04, etc.) which the expected values. The correct answer is "B" of course, so perfectly the second neuron should return 1.0, and all the others 0.0 – that would mean that the network had absolutely no doubts. So, let us calculate the distance ("how far from the perfection" the tool is) by summing up all the differences per each neuron:

Answer	Actual neuron output	Expected perfect output	Error rate per neuron
"A"	0.00	0.00	0.00
"B"	0.93	1.00	0.07
"C"	0.04	0.00	0.04
"D"	0.24	0.00	0.24
"E"	0.11	0.00	0.11
		SUM:	0.46

So, the overall error rate is 0.46 – this is not much if we realise that it could be as bad as 5.0 at worst. In addition, it is worth reminding ourselves that although we try to reduce the value of an error rate during the learning process (so a neural network learns to distinguish between defined categories better) we do not wish to push to reach 0.00 at all costs. The point of absolute zero may look tempting but it also increases the risk of the over-fitting effect – the network that perfectly matches all of the learning set answers, by doing it too precisely, may lose the ability to generalise tasks, to look wider and to resolve more difficult problems.

With the error rate calculated, backpropagation algorithms use this value as a parameter in weight updating formulas. The transformations are often quite complex and apply advanced maths theorems and that

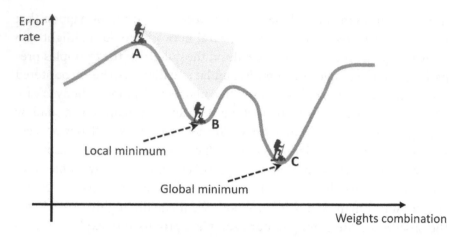

FIGURE 3.14   The error rate function illustration.

is why we will not focus on them in detail. Many of these formulas are already implemented as freeware libraries and can be successfully used as black boxes. Still, it is valuable to understand the common, high level concept hidden behind them. The main idea is to analyse an error rate not as a single value but as a function (of weights). Because an artificial neural network is usually quite a complex structure, there are many weights to be smartly changed (each connection between neurons by default has some weight assigned which describes its importance, capacity or width if we compare the power of signal with a water pipeline). That means the error rate function has many arguments which could be imagined as a multidimensional uneven carpet with bundles and concavities. Of course, anything of more than three dimensions (which we can see as height, width and depth) is quite difficult to imagine for people (except topologists* maybe). As a consequence, we rarely try to illustrate this function by focusing more on mathematical transformation itself. In this book, let us consider a single dimension for weights – then such a simplified error rate function can be drawn in a classic system of coordinates (see Figure 3.14). Each point of the line describes the level of error (the distance from a perfect answer) made by our application for each weight's configuration (here just one weight). So what is the next step? If we define the function correctly, the learning process becomes a mathematical task to find

---

* Topology is a branch of mathematics which is an extension of traditional geometry. Topologists often analyse properties of multidimensional objects. There is a joke among mathematicians that when a topologist is asked to investigate a square he prefers to extend this to an infinite number of dimensions just to see it better.

its minimum (point C in the figure). You can imagine this as a mountain hiking challenge – amateur tourists are asked to identify the lowest place in the mountain range, where the special prize is hidden. So, they walk across the area looking around for locations and positions of lower attitude, and when they notice one they try to get there in the shortest possible way. That is exactly how **gradient descent** optimisation algorithms work – a gradient is a direction of fastest function decrease (in other words the steepest path) calculated at each point and this direction is chosen by the system to go further in order to find the minimum. Just be careful. In this challenge, it is important to behave smartly and try to look wider because depending on the situation, going only by the algorithm may lead you to one of the local minimums (local mountain hollows not being the actual lowest place in the range). When we discuss the artificial intelligence learning process a local minimum may refer to the effect of over-fitting, which means that the network perfectly matches all of the small learning set answers but at the same time, by doing it too precisely, it loses the ability to generalise tasks – a neural network somehow gets stuck in the local minimum and becomes unable to learn anything more (hiker B on the illustration). The simplest idea to avoid such a situation or unblock the network getting stuck is to present to the network a greater variety of examples during the learning process. The examples should not be too similar so as to let the network look wider – the bigger the variations, the better the perspective. In the mountain analogy, despite the temptation to go down the hill as soon as possible (because the award is waiting and the other participants are on their way too), it is worth acting against the first thought and climbing higher (hiker A in Figure 3.14). From the top, one can see more* and the identification of the lowest valley in the range is much easier. Apart from a larger number of learning cases it is sometimes also worth randomising the weights a little during the learning process to make error rate reduction a little bit slower. So instead of a rapid gradient descent solution leading the system down the hill as fast as possible it is often better to jump across various weight configurations and then by reducing the jump length get closer to the global minimum. This kind of method is called **simulated annealing** and you can imagine

---

* The concept of the big picture (also known as helicopter view) is quite common in the area of modern business as well as software design and development. It is worth stopping your current work from time to time and considering the wider perspective of the work you do: who is the implemented application for? What is actually needed for your report? Where does the money come from? Even if it is your boss's role to monitor it, this knowledge helps to do your work better.

it as the spilling of a bag of bouncy balls above the curve of the error rate function (as shown in Figure 3.14). As the balls hit the graph, they start to bounce, initially quite rapidly and with large height and distance. But, with time, the bouncing intensity reduces and finally at the end of the process (when everything stops) most of the balls can be found in the deepest hollow – the global minimum. You can perform this kind of experiment on your own – just build some irregular surfaces using some home materials (pieces of wood covered with old carpet or plastic; or just hard pillows and a sheet in the simplest case) and some balls (bouncy, table tennis, even balloons – all depending on the surface and the size). One more thing – the algorithm's name is not accidental –annealing is a physical industrial process where metal or glass is heated and then allowed to cool down slowly to remove all the internal stresses. Metal or glass artefacts are then toughened – stronger, more durable and less susceptible to damage. Similarly, simulated annealing allows the neural network to learn better and become more accurate.

DEEP THOUGHT

"Artificial intelligence" is one of the most repeated technological phrases nowadays – you can hear it in popular TV talk shows as well as IT and scientific conferences. This is because the AI boom truly seems unstoppable and, at the very same time, newer and more breathtaking applications make us both excited about new hopes and opportunities ahead but also fearful of potential threats. As with everything new and rapidly developing in the history of our civilisation, AI has probably an equal number of fans and haters. The future will show who is more right in his arguments today. AI is a kind of black box for most people. For those who understand the topic a little bit more, another buzz words is what drives their imagination – deep learning – which is being used in various contexts and smartly included in many companies and product names. Just Google it to find that it returns more than 700 million results, three times more than "artificial intelligence" itself.

Deep learning is an artificial neural network technique characterised by the structure containing huge number of neurons, counted in tens or hundreds of thousands. You may start wondering how the algorithms have changed within the last few years to make so incredible progress from just university scientists' toys to devices and applications that realistically change the world we live in. And the answer may be quite surprising – the methods used have not changed much – in particular, in the most famous deep learning applications, you will still find a variation of the backpropagation weight configuration strategy. Its main ideas are exactly the same as when it was firstly introduced in the 1960s. So, again, what made it so incredible within recent times? The key change here is the scale. The biggest problem and progress blocker in the area of artificial neural networks was efficiency. Computers were unable to process too many weight updates too quickly – it was quite common that a pattern recognition learning process lasted weeks. Imagine something was wrong and another round was required – another few weeks gone. And because time means money and also fast feedback increases motivation, these kind of solutions were staying outside of the main streams of commercial technologies with limited resources and funds. Then, the era of **cloud** computing came and suddenly it was realised that you do not need a massive mainframe hangar-like device and tons of storage disks to calculate things quickly*. Within a few years, learning algorithm execution time was reduced from weeks to minutes and suddenly everyone on the planet has become able to do implement his own AI. And wide access means many applications and doable business plans. AI became a modern goldmine – it is where the money is and the biggest players agree that they will invest millions to potentially soon billions into it.

One of the most famous examples of successful deep learning applications is surely AlphaGo – a system that surprised even futurists, and made all other realise the world would no longer be the same. One of the oldest games still played, a magic one, and a fortress of complexity of humankind was finally captured. By defeating the strongest among men, machines showed that nothing can be really said as unable to be automated. To create AlphaGo, DeepMind architects combined the computing power of cloud solutions with some advanced neural network structures. The exact details can be easily found over the Internet and thus I do not think this technical information is necessary for the reader. Still, there are some interesting aspects in the solution that are worth highlighting and are related to the system's learning strategy. Unlike chess, to be a good (human) Go player, you do not need a very good (perfectly photographic

---

* In cloud solutions, the classic house-size computer is replaced by virtual machines and other resources distributed across the Internet. Anyone is nowadays able to, relatively cheaply, buy exactly the number of virtual machines, processors and memory space that he needs. It becomes available only to that user almost in single click. Resources that were available to the biggest companies only (that had money to buy a huge mainframe) are now open to smaller and smaller clients, as prices are decreasing every month.

one) memory to remember hundreds of sequences and opening strategies. You do not need a mathematical background (in chess useful for calculating position advantages). In fact, while playing Go, the most important skills are imagination and the ability to analyse a situation globally to correctly interpret the bigger picture. That is why this game is so popular and sometimes compared to ancient military battlefield analysis – the understanding of the situation as a whole is crucial in both. Focusing on a small part only may result in small wins of single troops but a huge defeat of the army as a whole. And as on the battlefield, the important skill is the ability to decode your enemy's action plan. World history has shown that armies, even the strongest ones (based on soldiers and cannons counts), have been unexpectedly destroyed by much weaker but brightly commanded forces. So, what the AlphaGo creators initially decided to teach the artificial neural network was the prediction of an opponent's next move. The learning set was composed from the master level Go play recordings kept in the archives of various events (it is worth comparing this with usual chess classes where students also analyse famous plays, especially the ones for the world championship). At the end of that stage, AlphaGo was able do it in 57% of cases. It does not seem much you can say – trying one hundred times the system makes 43 mistakes on average – quite far from the result requested to get a scholarship for academic results… But surprisingly it was more than enough to jump to an even more interesting phase (so remember –school marks are not everything – the way you use your knowledge is much more important than recalling learned facts on demand).

The second step was truly innovative and changed the way we look at AI. After the learning process was completed (achieving 57% of effectiveness in the prediction of an opponent's moves), the programme was distributed within a cloud and configured to play against itself (copies of its own instances). Each game improved the skills of the core. What is unique and fascinating, this technique is more similar to human behaviour and interactions than to what we usually associate with software engineering technologies. However, this shows something even more thrilling – it is a clear and incontestable proof of a quite old science-fiction concept saying that after achieving some level of competency, future computers will be able to improve their skills on their own, without any human help needed. They could learn extremely quickly and potentially forever (unlike humans, whose learning skills are reduced due to tiredness and ageing), leaving even the smartest among humans far behind. That may mean that strong artificial intelligence could be our last achievement. Everything invented later would be invented, designed and created fully and only by machines, without our suggestions, support or even approval. And at some stage totally outside our understanding...

Artificial neural networks, and deep learning in particular, have an incredible number of actual and potential applications, usually connected with object classification and pattern recognition. This situation is not so surprising at all if we think of it more deeply for a while. Pattern identification is one of the most crucial features of human perception, cognition and reasoning. It helps us to quickly recognise friends and enemies, situations that may be comfortable and, on the other hand, risky. The whole human learning process and, as a consequence, almost all of the skills and abilities we master during our lifetime, constantly improve due to the never-ending modification of the complex neural network happening day and night behind our eyes. The process that we are unable to control but which controls everything we do, feel and think. So, again it is not surprising that artificial neural network-based systems which imitate (on a small scale of course, at least nowadays) our brains' structure are so popular. The fundamental goal of IT in general is to create systems which automate human work to make our everyday tasks simpler and more comfortable and to let us skip all the (usually repetitive) activities that we do not like performing anymore. Deep learning is a perfect answer to that challenge as it combines machine precision with some human-specific abilities. That is why we will sure hear more and more breathtaking news from that area. And it is much easier to list dozens of successful application of these technologies than to look for an area of life that may stay free of it in future. AI tries to imitates man and we can expect that one day it will be able to do anything that we can faster, cheaper and more precisely.

Just to give few examples of the artificial neural networks applications, although you probably know even more at this moment. Deep learning is successfully applied in printed and handwritten character recognition (so-called OCR). It is used for picture analysis (so we can login to a PC or unblock a mobile phone just by pointing the camera at our face) and object identification within pictures or video clips (what helps to recognise dangerous tools or situation in city webcam monitoring). It also supports doctors in medical diagnostics, for example while reviewing X-ray images or matching symptoms with appropriate illnesses. Deep learning methods allow a car to drive a street smoothly without a driver or remote controller. Crude oil exploration, speech and voice recognition, airport luggage security scanning... all of these are just the tip of the iceberg. And for whoever does not yet know it, it is high time to understand this: we are a step away from the revolution. And within a decade our world will be totally different than nowadays.

✐ **NOTES**

- Artificial neural networks are AI solutions inspired by the structure of human brain. Still, the scale between the two remains incomparable. There are thousands of artificial neurons in the implemented structures against 90 billion neuron cells.
- A single artificial neuron is a simple function that activates (changes state from 0 to 1) if a signal strong enough is given at its input. The activation may be just a switch jump or more smooth value change depending on the application.
- As in our brains, a single neuron has no significant meaning – thus artificial neurons (like the ones in our heads) connect to each other, forming a complicated network structure. Each connection is defined by a weight, which refers to the significance of this connection. The bigger the weight value, the more crucial for network skills the connection is.
- At the beginning, artificial neural networks are quite empty and useless. Their skills are improved during the learning process. On the high level, this process is based on presenting to a network a set of examples and expected answers (for each example). After some number of learning iterations (cycles), a network is able to answer the presented tasks and to identify patterns correctly, allowing the network to deal with challenges never seen before.
- Practice makes perfect. The better learning process, the higher skills of a network. The quality and diversity of the learning set (examples) is one of the most crucial aspects here. If you teach a kid to recognise a tree by showing only birch trees to him then later the small boy may not see an oak or an apple tree as a tree. This effect is called overfitting.
- 100% may mean perfection; however, computational perfection is not a human characteristic. Being too good at resolving particular puzzles only makes a player weaker in other brainteasers. Leaving some space for imperfection allows a network to generalise problems and identify more patterns. Similarly, in schools, learning definitions and algorithms by heart must be accompanied with creativity and usual task classes.
- On the lower technical level the learning process of an artificial neural network is realised by manipulating the connection weights. The weights of more important connections are increased while others are reduced. That is quite similar to the process that happens nonstop in our brains. When we learn new skills (like playing piano, solving Rubik cubes, skiing, etc.), the connections responsible for these abilities become

stronger, consume more energy and are characterised by higher brain waves. On the other hand, skills not being practiced regularly are slowly forgotten – the connections become weaker and weaker. You can see it clearly if you ever been learning some foreign language – all teachers are agreed that the most crucial factor in achieving fluent pronunciation and richness of vocabulary is regular usage.

- There are various algorithms for weight modification during the learning process, most based on error rate calculations. Gradient descent optimisation suggests changes according to the direction of fastest error rate function decrease (as walking down the hill following the steepest path). Simulated annealing is based on jumping across various weights configurations and then by reducing the jump length – getting closer to the global minimum (similar to glass or metal annealing industrial processes).
- In computers, everything is representable as a sequence of binary numbers. That is why the potential applications of artificial neural networks are uncountable. At this stage of technology the biggest limitation is our creativity.
- The most important thing to take note of: behind your eyes is one of the most complex network systems in the Universe. The world's biggest companies, with incredible budget and funds, are unable to create a working reproduction (able to share our competencies). It does not matter whether you believe in divine creation, the chaos theory of random modifications or the perfection of the evolutionary process. There is one thing certain and incontestable – the human brain is the highest level of natural architecture. The more you study our physiology or the implementation of artificial intelligence, the more you become fascinated, overwhelmed and tongue-tied by this construction. **Our brains are priceless miracles.**

✎ **YOUR NOTES**

CHAPTER 4

Genetic Algorithms: From the Galapagos Islands to a Computer-Composed Symphony

I T WAS 1836 WHEN a young British scientist completed his five-year-long voyage on HMS Beagle, during which he was working on an analysis of the geological structure of lands, ocean islands, coral reefs and atolls, especially theorising their origins and changes over time, and estimating future erosion. The widely repeated story says that it was especially the visit to the Galapagos Islands that lit the spark of an idea that revolutionised biology over 20 years later. During the stay on the islands, the whole crew was eating tortoise as a main meal and the young man noticed that the shape of tortoise shells was not always the same. The locals replied with an even more interesting fact – the slight variation of the shape specifically points to the origin – seeing the tortoise, the inhabitants were able to say precisely which island the animal had come from. Although he did not collect any samples (he brought back to England three examples of separate species of mockingbirds instead), he had never forgot that observation. In 1859, he published a book that changed forever the way we look at the world around us. The book challenged the foundations of biology that had been know so far. The theory was both simply explained and revolutionary. It is was criticised for decades (for example by the Church which perceived it as in opposition to the Bible) but also brought to the author immortal fame and a top place in current academic handbooks. The man was Charles Darwin and the book, entitled *On the Origin of Species*, described the concept of the theory of evolution. Darwin not only noticed that individuals in some analysed populations vary from one another (sometimes even very significantly) and that these features or variants are often inherited from their parents. What is far more important, he explained that all living organisms try to adopt to their environment as quickly as possible. The individuals that are less suited are perceived to be less attractive to mates and thus are less likely to pass their genes (inheritable traits) to following generations. Every population constantly modifies so that each next generation is better prepared to live in a specific environment, i.e. they can run faster, hunt more effectively (with stronger muscles or sharper teeth and claws), are more resistant to high temperature, low quantities of water (for examples in desert-like climates) or, more recently, to air pollution. The same rule works for all organisms living on planet Earth (but probably also elsewhere – wherever life exists), regardless of the size and complexity of these organisms. So, on one hand, we can look at the flu virus that modifies (mutates) every autumn to infect as many people as possible – it adapts quite quickly, making last year's vaccines ineffective. Something similar applies to bacteria – the more medicines we produce and ingest, the more resistant the bacteria become. That is why we are starting to hear about so-called "super-bacteria", resistant to all known types of antibiotics (which are probably used too often nowadays in cases where they are not actually required at all). Viruses and bacteria adapt well – each new generation requires newer and newer treatment substances and methods. But exactly the same regularity may be found in the macroscopic world. While watching any TV programme on the animal kingdom, we are often surprised by the amazing survival or hunting techniques (insects walking on water, chameleons able to change colour to match their surroundings, crocodiles with a jaw grip strength of over 2 tons per square centimetre and many more). Finally, we are not omitted – various human populations physically adapt to their current climate or other environmental factors. Recall inhabitants of the warm and extremely sunny areas of central Africa – the dark pigment in their skin helps to reduce the amount and intensity of sun burns. On the other hand, some research on Inuit people shows incredible adaptation to low temperatures – they can work outside building ice constructions without gloves – something that might cause significant injury (including frostbite) to others.

EVOLUTION SQUEEZED INTO MILLISECONDS

Evolution is a brilliant mechanism of nature, complex in detail but at the same time ingeniously simple in its assumptions. All living creatures strive for survival – this is the most fundamental instinct – to do anything possible to stay alive. A surrounded rat is aggressive and releases internal resources of such energy that it is sometimes able to successfully discourage even a group of two or three attacking cats. A fox trapped in a snare can bite off an entire limb just to get free. Survivors of sunken ships have been found alive (although extremely exhausted) weeks after the shipwreck. For the same reason, legends of the Fountain of Youth (a spring that restores youth to anyone who drinks or bathes in its waters) appeared in antiquity and the Age of Exploration when Columbus' discovery of the New World brought new hopes and expectations. Paradoxically, many actually devoted most of their lives unsuccessfully trying to find the true location of the Fountain... The desire for eternal life changed from an instinct to an obsession. Still, the deepest thought encoded in all brains is the same. Every living organism on Earth, from insects to elephants, does anything it can to extend its existence. To achieve it, it tries to acclimatise to the surrounding environment as much as possible. That implies constant and nonstop changes in order to gather more energy (to find and reach food easier by having great eyesight and smell and manipulating limbs; to breath efficiently at higher attitudes), better avoiding trouble (they can run away faster from predators, possess indestructible armour, have very sensitive hearing and eyesight) and finally looking for a heathy partner to make sure their genes don't vanish after death. That is what the technological race in the world of nature is all about.

Each entity in the world of nature strives for survival, safety, gene transmission and, if all the former requirements are already met, as comfortable life as possible. Adaptation to the surrounding environment is a key element to achieve these goals. Changes in body structure may be slow but are constant and more visible in each following generation. The history of life shows incredible application of modifications. For example, one of the most crucial, and spectacular as well, events in the history of the Earth was the vertebrate land invasion that started around 400 million years ago. You may wonder why I have just called this process a spectacular one. But think about this for a while and compare water-adapted fish and four-legged walking creatures. I would say it is much easier to say what makes them different than find some common features. Still, life evolution made a fish walk on land – something unexpected, even in the weirdest science fiction movies. Although the details are not fully agreed on by biologists, animals were strongly motivated to make that incredible step forward. Mainly because of slow changes in the marine environment that started to make it less and less comfortable for many species – there was less and less oxygen, combined with changes in temperature and increased salinity. In addition, the seas were becoming a home for more and more species – from an almost empty desert, they simply started to become overcrowded, which increased the chances of disease and, even more importantly, competition. More mouths and food requirements always makes it more difficult to find a nice meal for yourself. And, like people leave crowded cities and move to villages to find some space, calm and fresh air, similarly vertebrates were pushed out of the water to find new places to live. The barriers to transition were extremely huge (if the animals were people, they would probably never decide to do it themselves). Changes in senses were required – vision and sounds are totally different in the two environments, for example a fish taken out of the ocean is practically blind and deaf elsewhere. The placement of the eyes on head may sound obvious to us but try to ask the world's best surgeon to just move it on a fish's body and keep them working. He would probably laugh at you, listing dozens of issues in such an operation. There are also different pressures – a body must be characterised by different gas exchange and water balance. The body must also be more waterproof! Changes in muscles and bone anatomy are mandatory too – an animal needs to be able to walk in opposition to much more intensive effects of gravity. The same evolution that moved vertebrates from water to land is also responsible for the amazing variety and sizes of dinosaurs at the end of the Cretaceous period. That was the time of *Tyrannosaurus rex*, one of the largest land predators of all time, often called a perfect killing machine. The impact of a meteorite is the most probable reason for dinosaurs' extinction at the end of this period. If this had not happened, evolution would have certainly modified the species further. We would have been finding much bigger and even more scary skeletons nowadays. Or maybe we would have never had a chance to expand our civilisation to its current size if we shared the world with dragon-like creatures. But evolution works constantly, sometimes much less noticeably. It often helps in lifestyle adaptations and in optimising inherited behaviours. Although we often do not realise it, evolution was a factor in distinguishing some male and female skills. Of course it is not a rule and you can point at many examples against it, but statistics are clear – in individuals with Northern European ancestry, over 8% of men and only 0.4% of women experience congenital colour vision deficiency. Woman distinguish and identify colour much better and there is a reason for this. Traditionally, women took care of the home – they were responsible for plant collecting, cultivation and food preparing. It was crucial for survival to clearly distinguish between healthy and poisonous plants. On the other hand, men are usually quicker at recognising moving objects (what we sometimes refer to as reflexes) and have better orientation in the field. Both are crucial in hunting, to notice a beast in the wild and find a way home after days spent outside of a village...

What has not been strongly mentioned earlier is that the process of evolution is not a quick one. Of course, it depends on the average length of life of a particular species. But if you realise it needs tens, hundreds or thousands (depending on the mechanism being modified) of generations to make the change visible and useful. Regardless, the process of evolution in the animal kingdom can be always described as never-ending circles of life. The first step in every cycle is a natural selection. Each individual strives for survival and to pass its genes further. That is why an individual looks for a healthy, strong, and well-accommodated partner to ensure a safe life for its future family as well as numerous and healthy offspring. Individuals with expected features are more likely to become parents, while others, less well-adapted, often die alone, making their genes gone forever. As time goes by, the "weak" genes are slowly eliminated from the population. Yes, evolution is cruelly effective – whoever is not adapting well will sooner or later disappear from the history. Single genes and entities are sacrificed to make the population as a whole stronger. The better-adapted individuals copulate, giving the life to the next generation. And the process repeats. Forever. In the 1980s, the process of evolution started to inspire computer scientists* to build mechanisms for non-schematic analysis and automated actions. Mechanisms are free from stereotypes or bad habits (sometimes recognized as a first step towards artificial creativity). Although these techniques are successfully used to resolve even

---

* Notice that this is another example, after neural networks, where research in the area of biology has driven development in computer science. This may be surprising, as it is quite difficult to find more distant branches of science. Still it proves the importance of wide and non-schematic thinking – you never know where the huge discovery is hiding...

complex problems, the algorithm is quite simple. Like the biological inspiration, we start with a population of individuals which evolves in every generation, finally ending with an individual that perfectly meets our requirements. These methods bring the incredible power of evolution into IT applications while at the same time dealing with the biggest showstopper in the world of nature – time. Inside the computer we can fully control time, describe the environment, build our own populations and define selection rules. The entire process takes milliseconds instead of millions of years.

ARTIFICIAL DNA

As we already learned in Chapter 3, in the world of IT, everything is a number. Whatever you perceive as an end-user – text, an image, a video, a song, a live stream or virtual reality in a game – behind the scenes, deep in the device, all of these are long sequences of 1s and 0s. Similarly, any question or answer is also represented in the same way inside a machine, as well as the current state of all of the components. The numbers describe what our computer does, what it knows, what it remembers, and what other devices it can cooperate with. Numbers define the machine and make it unique. Knowing this, scientists found, again, an interesting analogy between the artificial and the natural world. How? Because we are all also defined by unique sequences of data. The difference is that these are not sequences of numbers stored on a hard drive but rather two chains of nucleotide molecules forming a beautiful art-like structure. This structure is called DNA and is stored in every cell of every living organism on the planet. It carries genetic instructions defining the process of growth, development, functioning and reproduction. In other words, it describes all the physical aspects of an organism, from the colour of the eyes to the strength of specific muscles and, on the other hand, inclination to specific diseases. Scientists decided to follow this analogy in solving complex computational problems. Let us recall quickly the knapsack problem described in Chapter 2. The knapsack's volume is limited and the weight cannot be too large since it has to be easily carried by the thief while escaping from the store he broke into. So which goods should he take? Is it better to grab two TV sets or three laptops, or maybe a laptop and four tablets? It all depends on their specific equivalent in dollars and the relationship between weight and a value. Finding the perfect combination of items proves to be a highly complicated task with no quick answer. To get a perfect answer, one would need to try all the variants. You may start to wonder what this has to do with DNA… Actually, any combination of the elements in the thief's bag, any potential result, can be represented as sequence of 0s and 1s:

1	0	1	1	0	0	1	0	1	1	0	1	0	1	0	0

Here, each position refers to a particular electronic item, e.g. first position: TV set number 1, second position: a TV set number 2, third position: a radio, fourth position: laptop number 1, etc. At the same time, the value in a specific position describes whether a specific object is put into the knapsack (then 1) or left in the store (then 0). So, in the example sequence above the first TV set is in the knapsack but the second one is not. This combination may be good or bad, but the important thing is that any combination of the answers can be described like this. Any question given to a machine (not necessarily the knapsack problem) can have answers described by this kind of sequence of 1s and 0s of a defined length. In artificial genetic algorithms, such a sequence is often called a **chromosome**. The cycles of life are then imitated within a computer, as shown in Figure 4.1. The whole process of finding the solution follows biological evolution: it all happens in cycles in which chromosomes are treated as living individuals – there is a pseudo-natural selection that identifies better and better results, crossover that replaces replication, mutation that simulates unexpected gene modifications, and the birth of a new generation. All in cycles repeating one after another until the optimal answer is found. Evolution implemented in a computer. And squeezed into milliseconds.

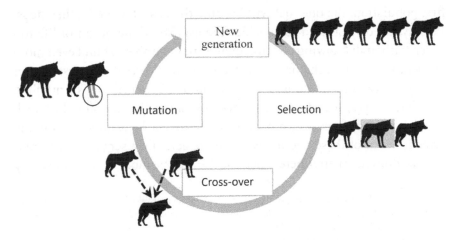

FIGURE 4.1   The cycle of a genetic algorithm.

THE BIRTH OF LIFE

The origin of life is a mystery that both inspires scientists and feeds the imaginations of ordinary people around the globe. Although some chemical processes and early-stage evolutionary steps seem to be quite well explained nowadays, the initial moment, the spark that started it all around 4 billion years ago, is still an open question. Some scientific analysis suggests that it is a lucky combination of climate conditions, slightly acidic water pH and intense UV light. In this very specific cocktail of parameters, the non-living molecules colliding with each other one day accidently formed the very first protein. Some believe it is God's wish that started it all at a very specific moment. There is also a hypothesis that the first living structure actually arrived to Earth from space inside frozen pieces of rocks falling on the planet's surface. That would mean we are all partially the aliens that we are looking for with telescopes. The state of the art in this domain is a little bit analogical to the cosmological macroscale. In the same way, we also know quite a lot about how the Big Bang looked and what was happening within the first nanoseconds following it. But again – what initiated it? And how did it happen if there were no time and space prior to it? On the origin of life, there is also the question of why this happened exactly here. On this third planet from the Sun, a very average star on the periphery of the Milky Way*, one among millions of galaxies in the Universe. Is life on the Earth something absolutely unique, or maybe a regularity quite common across space?

In computer-located genetic algorithms, we imitate the evolutionary process. And as in the natural process, we need to emulate the birth of the first population, the one that will become the basis for the further steps. Luckily, we do not need to consider the aspects of the origin of life too much (although knowing the answer, we might be able to build even more efficient artificial mechanisms). Instead, we should recall the observation we have already made earlier: in the world of, IT everything is a number. All files, images, sounds, music, videos or games that are stored on disk or in a cloud, a process by any application or displayed to a user on any screen – each of these and many more are actually a sequence of 0s and 1s. Size does not matter here – behind the scenes, modern machines only

---

* To realise how far from the centre we are, simply look in the night sky in some isolated area, free from city lights (so-called light pollution). You will see a lighter shade filled in with thousands of dot-like stars. That's the centre of the Milky Way. We really are a rarely visited, old-fashioned province. Full of monotony and having nothing to be proud of. But maybe the truth is exactly the opposite?

manipulate digits. Simple mathematical operations like addition or sub-traction change boring strings of numbers into everything we perceive as the magic of technology. So, as everything is a sequence of 0s and 1s, in the same way, any answer we wish to get from a computer can be repre-sented in such a form. Let us assume that the answer we are looking for is the knapsack problem. In a very first step of the algorithm, we emulate the birth of life, the creation of the initial population, a set of the very first entities of a specific kind. So how to create life? Just follow the biolo-gists' hypothesis of the lucky guess that made us all end up where we are now. An accidental combination of factors that started it all. In genetic algorithms, we already know the length of the sequence (e.g. each element in a sequence refers to a specific item in the knapsack belonging to the thief who breaks inside a store), we just need to find the optimal value of every position. To do this, we choose the initial individuals (solutions) absolutely randomly. They could look like this:

1	0	1	1	0	0	1	0	1	1	0	1	0	1	0	0
1	0	1	1	1	0	1	0	0	1	0	1	0	0	1	1
0	0	0	1	1	1	1	1	0	1	0	1	0	1	1	0

. . .

0	1	0	0	0	0	1	0	0	1	0	1	1	1	1	1

We need to choose the size of the population, which usually depends on the length and the complexity of the problem we wish to resolve. Here, let us assume that the population size is 100 – we randomly choose 100 sequences of 0s and 1s. To do it in a proper way we would probably need to throw a coin $100 \times 16 = 1,600$ times. The good thing is that computers have random number generators built-in – it would take microseconds to get the starting population. Just blink quickly and we are done with the first step of a genetic algorithm.

NATURAL SELECTION

Having a **population**, a collection of individuals represented by chromo-somes, we can now properly start the process of artificial evolution. To do it, let us think again of the biological inspiration – the key element is natu-ral selection – individuals that are stronger, faster, bigger and, generally speaking, better adapted to the surrounding environment are immediately perceived as more attractive and interesting potential partners. This helps positive genes to be passed to the next generation and so to give the new-comers a higher chance of survival. It is easily noticeable when observing animals living in herds with a single dominant which is usually the parent to most of the young in the group. In the other corner is the omega animal, treated as belonging to the worst category and the least valuable member of the family. The strong and powerful alpha shares his genes widely while the omega might stay outside forever without any chances to pass its genetic code to the next generations. All of the features that suggest better environmental fitness are desired and thus mainly recognised as attractive to representatives of the opposite sex. You might think it is a primitive mechanism that can be found only in less developed species. But this is not true. These elements even affect humans partially (luckily not fully, as we still use higher feelings), influencing our decisions when building new relations with others. Just think of the popular symbols of being sexy. Of course we could say that it is not a body that makes the final wedding decision but it sure is what we look at first when meeting someone for the very first time. Before exchanging a single word, we can already say whether a particular person is attractive to us. Men are usually attracted by women's full breasts and rounded bottoms, which actually means a potential partner will be able to feed infants quickly and a birth should happen without any complications. On the other side, tall athletic looking boys usually have a higher chance to find someone to go out with. Why? A strong man is recognisable to the one that, in case of danger, would be able to protect his partner and kids. A taller one is also perceived as more dominant and thus potentially ensures his future family's better position in a group. That means that these features put the individuals in a more comfortable situation in an everyday environment. And better adaptation brings attraction. Despite all the self-development made, we are still partially driven by these natural instincts that grew from the fundamental mechanism of natural selection. For the same reason, faces and bodies that are perfectly symmetrical are named icons of beauty, and deformations, even the smallest, can affect the overall perception of a person. Why? Because symmetry is a synonym of biological balance and so health, while defects suggest illness, bad habits or risky lifestyles.

So, we are clear on how natural selection works. Now how do we switch to an artificial one? What makes one chromosome more attractive than another? In the case of genetic algorithms, individuals are simply possible solutions to a given problem. At the same time, the environment (which chromosomes tries to adopt to) is represented by the so-called **fitness**

function which for each individual resolution returns the information on how good it is at adapting. For example, if we want to design an integrated circuit, an initial individual would be a random circuit schema while a fitness function would return the performance statistics of that solution. Going back to the example with the thief who has just broken into a store: the chromosomes (sequences of digits) describe the potential ways of packing the knapsack and the fitness functions could represent the value of the bag. The more expensive the stuff inside (altogether), the better the solution, so the better environmental fitness. So, in our case, for each of the 100 chromosomes initially generated (randomly), we calculate the total value of items chosen, and this value becomes the fitness function related to a particular individual. We should also skip all the sequences where the sum of the weights exceeds the capacity of the bag – even if these elements are extremely expensive, the thief is unable to leave the store with it and actually gets nothing. So, to skip these cases, we can simply assign 0 to them (zero has no benefit at all). Now, having the value of the fitness function assigned to every chromosome we can order them by this – from most to least valuable combination of stolen items:

Chromosomes:																Fitness value:
1	0	1	1	0	0	1	0	1	1	0	1	0	1	0	0	$1,234
0	1	0	0	0	0	1	0	0	1	0	1	1	1	1	1	$932
0	0	0	1	1	1	1	1	0	1	0	1	0	1	1	0	$722
1	0	1	1	1	0	1	0	0	1	0	1	0	0	1	1	$712

. . .

The next step refers to the idea of the natural selection directly – we simply choose the top ten chromosomes (best-adapted) and drop everything else. We stay with a small subset of the best solutions among the generated ones.

CROSSOVER: A NEW GENERATION BUILDS A NEW WORLD

Genetic algorithms work as heuristics – they might not identify the best possible solution but can still find an answer that is good enough for our applications. Like humans who do not need to calculate their position with the precision of fractions of a centimetre to walk down a street, a machine does not need use massive amounts of energy to find the perfect variant. If we want to design an integrated circuit, we can give a threshold performance value that we want the circuit to meet. So, as we have already found the ten best answers (in the previous section), we may spend a while to check whether perhaps the top one is what we expected. In such a case, the algorithm could be stopped – we are pretty lucky find quite a nice solution just by random tries (make sure to play the lottery then too!). Otherwise, it is time for **crossover** – connecting partners in order to give birth to new generations of genes that are a combination of the ones inherited (usually in equal proportion) from each of the parents. We work on ten sequences of 16 elements (0s or 1s) each. To emulate the gene transmission process, simply cut each chromosome into two halves and combine the left-hand side parts with all of the other prepared right-hand side pieces. Combining them again into a single chromosome of 16 items we get totally new variants with some features inherited from each parent. In the simplest version of genetic algorithms, the handover is achieved via simple string updates as in the example below:

1	0	1	1	0	0	1	0	1	1	0	1	0	1	0	0	Parent I
								+								
0	1	0	0	0	0	1	0	0	1	0	1	1	1	1	1	Parent II
								=								
1	0	1	1	1	0	1	0	0	1	0	1	1	1	1	1	Child

As we split each of the selected entities into two items and combine them with the remaining elements of the rest. That would result in creation 10 × 9 = 90 entities which would form a fresh new population of 100 (also including the ten chromosomes initially selected). As the new population is created based on highly ranked individuals in the old generation, each population achieves (on average) better scores than the previous one. Simply speaking, on average children exceed their parents in the skills that are useful from the survival point of view: they are taller, stronger and more resistant to mass infections. Of course, there are many more external factors that affect life expectancy in a particular population, like military conflicts, access to medical care and technology, pollution, local and global pandemics and many more. Coming back to our genetic algorithm, each next generation contains chromosomes (sequences of digits) conferring more and more accurate results. Still, we need to remember that these are heuristic techniques – as in real-world evolution, there is no guarantee of when we will achieve the perfect individual (if at all). Thus, before we actually start the algorithm, we need to understand well what we want to achieve, where the value that satisfies us enough is. In other words, although we may not get the best possible answer, we will probably find a result that is more than enough for our applications. To give an example: if we want to calculate the distances between stars across the Universe, we do need to focus on inches, we do not even need to think of miles or thousands of miles – light-years* are pretty much enough for most amateur astronomical purposes. One may say that such a low precision is a disadvantage to such AI mechanisms, but the truth is we are unprecise too. And paradoxically that is what lets us live, work and have fun effectively. Similarly, we need to check if any of the chromosomes generated by the genetic algorithm are good enough for us. If so, we stop the algorithm. Otherwise we continue – selection (based on the fitness function) is performed on the newly created generation, and then crossover is done again to get further, even newer populations.

As we have already said, in the simplest version of the crossover, we choose ten of 100 elements (the best scored chromosomes), then split each into two equal halves, and combine each of the left-hand semi-chromosomes with all the right-hand pieces. Is this method an optimal one? Actually, it depends on the application we are considering and the precision we wish to achieve at some stage. One of the important factors is of course time – we want our algorithm to work quickly so that we get some valuable results in the smallest possible number of iterations. The lower the number of generations that need be generated, the faster the calculations. So what can we do to help the evolution proceed faster? We can manipulate many parameters of the selection and crossover mechanisms – below you can find three examples.

Splitting strategy: Here, we split each chromosome into two equal halves (each with eight digits of the initial 16); however, we may consider different proportions of cut depending on actual applications. We can also try to change these proportions in time, so for example in the first generation splitting chromosomes into halves and in later ones, when the evolution mechanism is warmed-up, having the left-hand side piece bigger than the right-hand side one. Such a strategy can help to stabilize some parts of the chromosomes after a particular number of generations. So, looking for some analogy we can say that since the solution is partially found, we may want to keep some bigger part of the resolution sequence (chromosome) unchanged. A little bit like when you create puzzles – having one part

---

* A light-year is a length unit used in astronomy, defined as the distance that light travels (in a vacuum – encountering absolutely no obstacles on its way) in one year's time. 1 *ly* is around 5.9 trillion (5,900,000,000,000) miles. The nearest star (other than the Sun) is Proxima Centauri, about 4.22 *ly* away.

of the expected illustration already completed, we do not want to change anything in that part. Another idea is to randomly choose the place to cut (the point where we place the scissors is randomly chosen for each new population). So which way should we go? There is no general recipe here – it all depends on the problem to be solved, and some experiments are often performed first to see how it goes. In addition, it is worth staying balanced if we're not really sure which option to choose. Why? For example, if we split our 16-bit-long chromosome in a proportion 14:2, then we quickly realize that the right-hand side part with a length of 2 bits can have only 4 possible options, as seen below:

0	1	0	0	0	0	1	0	0	1	0	1	1	1	0	0	Parent I
														0	1	Parent II
														1	0	Parent III
														1	1	Parent IV

That would mean that such a proportion immediately reduces the number of possible combinations (from 512 for an 8-bit-long piece) and so the chance to find an answer unfortunately drops quickly. Just like in a crossword puzzle – if you fill in some random letters at very beginning, your chance of completing it is rather low (unless you are really lucky or have some sixth sense).

Size of the population: In our example, the size equals 100, but generally speaking, the more individuals in each population the better. Why? Because as the initial chromosomes are randomly chosen, more examples bring a better chance to find better ones. That also means more combinations generated in the crossover process. All of this suggests that the expected individual would be found earlier. Just stay reasonable. Too large a population may significantly extend the time needed for selection and crossover, and potentially mitigate all the benefits we gained by limiting the number of iterations needed. In other words, if you take care of wildlife in nature, huge populations and herds of several thousand individuals give you great diversity and opportunities to observe huge-scale interactions, and the chance to find some incredible individuals (it is not surprising that many biologists spend their lives in jungles). But the quid pro quo is that such a huge population is difficult to control, and the natural selection is almost impossible to trace. So, as quite often, balance is the key.

Size of the selection: This one is quite interesting too. In our example, we chose the top ten (best-fitted chromosomes) in each population (of 100 entities). But is this the best solution? Some time ago, I performed a small experiment implanting multiple variants of the same genetic algorithm supposed to resolve an example knapsack problem. Surprisingly, the answer was found much earlier when I increased the size of the selected group (to be crossed over later) to about 40% of the initial population. So, again, it is important to keep the balance. Selecting too many entities (e.g. almost all of the population) for the next phase may slow down the whole evolution a lot – it reduces the influence of the fitness function and the adaptation needs that are not important enough are likely to be ignored in the world of biology. On the other hand, and what is even more fascinating, if we select too few items (e.g. just the top five), further crossover will not improve the chromosomes quick enough. In other words, if you have just a few close-to-perfect individuals, it is difficult to mix them in such a way that the next level is achieved. They are just too similar and too optimized already to be easily modified. This is also observed in genetics – mixing populations and greater biodiversity speed up evolutionary changes. Small, isolated populations of plants or animals located on distant islands are often similar to the ancestors of similar species in other parts of the world. Isolation and small groups make the changes progress faster in the beginning, but progress dramatically slows down without any occasional fresh input from outside. If we recall Chapter 3, we can see that this problem is analogical to the case of local minimum in the artificial neural network mechanism: the algorithm itself does not know whether it moves in the direction of the global or the local minimum. After falling into the local one once, some extra action needs to be taken to get out of there and proceed further. You can extend this observation even to the areas of sociology and science: it is quite often that isolated communities live in the same, quite primitive from our perspective, way as hundreds of years ago. Companies start to develop faster when new and more ambitious people are hired, people that are open enough to point at some repeated mistakes (things done in a certain way for years) and are also happy to share their fresh view and lightning ideas.

Evolution needs some space.

The mechanisms of selection and crossover are both easily understood on a high level, but at the same time are challenging in the aspect of technical details. Exactly like in biology, various topics must be considered. Recent progress in GMO (genetically modified organism) production, genetic modified food, genetic-based medical therapies... all this shows that this area will soon be explored even further. And that may generate some new concepts for computer-controlled genetic algorithms as well. In both cases, though, we need to remember one thing: it is difficult to guess and predict all of the consequences of our modifications. It is not simple playing with bricks. That actually reminds me of an old anecdote often associated with one of the most famous Irish writers, George Bernard Shaw. He once got a proposition from a very pretty woman to become the father of her children.

- "Just imagine", she said enthusiastically, being absolutely serious. "Inheriting your intellect and my body, they would be simply perfect."

- "Yes, with pleasure", Shaw replied. "Just I'm afraid of the inheritance working the opposite way!"

Evolution likes to follow its own paths, paths that are difficult to notice at the first sight – like small trails in a rainforest.

X-MEN AMONG US

Mutations are nothing more than mistakes or, we could say, unplanned modifications in chromosomes. Usually, they are found to be harmful, causing serious disorders like Down syndrome (an occurrence of an unwanted copy of chromosome 21), accelerated aging (where some systems or organs of one's body age prematurely) and, most often, various types of cancer. However, there are also examples of beneficial mutations in live species around the world. And although the chance to meet Magneto of the Marvel® universe (for those who do not know – this evil comic character can control magnetism and so move and modify metal objects without touching them) next door is pretty impossible, some mutations are quite impressive. For example, there are modifications found among some European citizen populations that make particular people resistant to the HIV virus and thus protected from or at least which can delay the progress of AIDS. The interesting aspect is that this mutation may find its origin in the 14th century when Europe was being ravaged by the Black Death. The mutation helped save some lives and, as it was crucial for survival, started to be transferred to following generations. So, paradoxically, the micro-residue of this forgotten medieval disease might help us to fight the AIDS pandemic in Africa. Mutations do not only occur in humans, of course, but are common elements of the global natural process affecting all living things. And some permutations beneficial for particular species do not need to be good for us in the end, like in the case of bacteria which develop antibiotic resistance.

Mutations can be introduced at the very early stage of an organism's growth. Still others can occur during the lifetime as a result of environmental influence. Toxins contained in car exhausts or cigarette smoke as well in the ones hidden in strong alcoholic drinks or some fast foods may lead do some unexpected changes on the cellular level. Exposure for a longer time to the above can cause serious harm to an organism. What is even worse, some external factors may generate mutations even in a single, short-time contact, e.g. UV (a long period of sunbathing can damage the skin irreversibly) or ionizing radiation (like that which accompanied nuclear plant failures in Chernobyl, and more recently in Fukushima, Japan). Finally, some mutations are spontaneous and their origins are difficult to trace – it is likely that some random changes are programmed deeply in the nature evolution process to speed it up in some cases and help species to adapt faster to a quickly changing world. Of course, random, spontaneous modifications sometimes lead to failures as well. It is just that in the final calculation, the benefits outweigh the losses. And the individuals affected by harmful mutations are quickly eliminated in the natural selection. Mother Nature is cruelly efficient.

In artificial IT applications where we simulate the evolutionary process to solve some technical problems or answer complex questions, we can also successfully apply mutations. And, in fact, it is quite a basic case to programme – it is enough to randomly choose one of the cells of the chromosome (i.e. a field in the string) and change its value to the opposite. So if it was 0, now it will become 1, and on the other hand, 1 is converted to 0. As simple as that – see the example here:

1	0	1	1	1	0	1	0	0	1	0	1	1	1	1	1	Original
													↓			
0	1	0	0	0	0	1	0	0	1	0	1	1	0	1	1	Mutated

A mutation is the final step in a genetic algorithm. Now that we know them all, we are ready to put all the pieces of the puzzle together.

EVOLUTION OF A SOLUTION

As we have already said, genetic algorithms in modern IT imitate the processes that have happened around us for millions of years. The processes that let various species adapt to the changing environment, forming extraordinary skills as well as organs and tissues that help them to hunt or to defend themselves. If we think of it for a while, these effects of arduous but constant progression could often exceed Hollywood script writers' imagination: dogs that can trace a person based on just a single smell molecule left in the air, a spider's web so strong that inspires Kevlar bulletproof vests, blind bats that are able to fly incredibly fast across cave labyrinths using echolocation known well from ultrasound examinations and submarine navigation.

The world of animals and plants is extremely complex and still surprises scientists. It is not certain if we will ever understand all of the dependencies that form its foundations. And it is all based on the sequence of pretty simple mechanisms that we discussed earlier. So, it is not surprising that they have become a basis for one of the most efficient AI techniques. So, to summarize once more, let us describe together how the genetic algorithm mainly works. We do it in two phases. The first is the preparation, so all of the things that need to happen before actually we turn on the computer (so it is a design made by a developer). The second stage is the execution, so the actual sequence of steps performed by a computer, extremely quickly, to provide the output to a user.

Preparation (a man):

1. Note down the solution you would like your AI to find.

2. Define the solution (problem) using the form of a chromosome (so a sequence of 0s and 1s of a particular length – you need to agree which position in an array means what).

3. Choose the fitness function, i.e. the value that will tell you that the solution being looked for has been found (maybe not the most perfect one but just enough for your needs).

4. Implement the computer programme to be executed.

EXECUTION (a machine):

1. Create an initial population (a set of random chromosomes), e.g. 100 items.

2. Apply the fitness function to order the population (best chromosomes at the top).

3. Check if the top chromosome is already enough (fitness function above some defined threshold). If so – go to step 8.

4. Choose the top ten items and split them into halves.

5. Combine each of the ten left-hand-side halves with the ten right-hand-side halves.

6. The new population is born ($10 \times 10 = 100$ items again).

7. Go to step 2.

8. Print the top chromosome sequence (the answer to a user).

This AI mechanism of course requires some programming skills to implement the application (to programme the machine to follow the execution steps). Still, you can try to imitate it manually yourself even with zero knowledge of how computer works. Just prepare a set of 80 equal coins in a paper bag. Let us assume our goal is to get as many heads as possible in a sequence of eight coins. That is our target. So, take eight coins from the bag and flip them one by one, forming an ordered sequence from them, as shown in Figure 4.2.

Now repeat the same action for the rest of the coins from the bag, getting ten lines at the very end. These are your chromosomes. The task we agreed to give to our algorithm was to get as many heads as possible, which means our fitness function should answer the question of how close we are to the perfect result. In our case, this function can simply be the number of heads in a sequence (where eight is the perfect result). In the example above, the value of our fitness function equals three. Now calculate the function for the rest of the sequences. Once this is done, the time for selection comes. The process is rather straightforward: keep the top sequence unchanged and for the three best sequences (with highest fitness function value), split them into halves and combine all pair to get $3 \times 3 + 1$ chromosomes (Figure 4.3).

You can add some random mutations too (by turning one of the heads into a tails, or vice versa), just be honest! – make sure your mutation is truly random – perfectly, do it with your eyes closed. The young, better, new population is now born. First, check whether maybe one of sequences already contains eight heads. If so, you can stop the algorithm – the result has been found. Otherwise, calculate the fitness function, choose three

FIGURE 4.2 The chromosome-like sequence of eight coins.

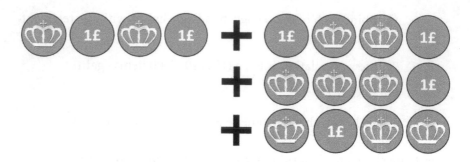

FIGURE 4.3 The crossover of coins chromosome.

best sequences, etc… You can count how many populations have to pass away to get to the result – it should not be many of them, but you never really know. If you like that game, you can try again with different parameters to check if you are able to reach the result faster. How? Add more or less mutations, select four or just two sequences for reproduction, split the chromosomes differently than 4–4, e.g. 3–5. You did it – you have performed the world's most fascinating and powerful process using 80 coins and the knowledge you have just learned.

I hope you enjoyed the experiment above – when we think more deeply about the algorithm used, one can realise quite quickly that the process itself is not solely the domain of Mother Nature and computer scientists. Imagine a master chef who wishes to create an absolutely spectacular and unique dish. To get wider recognition and, with some piece of luck, even a Michelin Star, preparing a typical meal, even a very tasty one, might not enough. It is important not only to manipulate herbs and spices but also to combine different and unobvious flavours and aromas. These variants seem quite strange and unconvincing until you try it. And when you try, you fall in love with it. That is the secret: unexpected combinations and hidden ingredients. Mixing two popular dishes into one unusual course is often the key to the success – can you see the analogy with selection (popular dishes) and crossover (mixing elements of both dishes) of the genetic algorithm? Not surprising – lots of ideas and world-changing innovations are based on similar research strategies: connect existing items in a non-existing way to create new things we would never have thought of before – and generating new needs as well as pushing development and technology in totally new directions, down some unvisited paths. Creating objects our world would not be the same without: writing (voice + drawing), ceramics (stone + glass), ice creams (ice + cream), cars (cart + engine), cinema (picture + movement), Microsoft Windows (programming console + visualisation), smartphones (computer + telephone), and social media (Internet + club).

Our society, fashion and technology follow the same evolution algorithm as the world of nature.

EVOLUTION IN IT

Genetic algorithms already have so many various applications that the best way to look at them closer is to divide them into three general categories. The first group contains applications that are created to find solutions or answers to some nontrivial questions. As the algorithms analyse hundreds of thousands of populations in a second, they are able to combine and review more variants during one's coffee break than a scientist would during a years of research. The algorithms help find a solution to the **knapsack problem** (which we already discussed earlier), are used in computer games (so our artificial opponents become smarter and more challenging enemies on a virtual battlefield) or in DNA structure prediction (valuable in laboratories where newer, more efficient medicines are produced).

All of this sounds really powerful, but finding such solutions is just one of the areas of application – the techniques of genetic algorithms are even more popular in design. However futuristic this may sound, nowadays many tools or mechanisms are not only used by computers but also designed by them. Yes, machines design and build machines! It is here and now in the second decade of the 21st century, not science-fiction anymore. Algorithms propose aerodynamic shapes for vehicles, analyse complicated projects or design electric circuits. Look at Figure 4.4, which may look uninteresting at first glance. It changes as soon as you realise that it is a NASA spacecraft antenna and that this complicated shape was fully designed by an evolutionary algorithm to optimise radiation parameters. Perfect space communication due to a strange uneven shape suggested by a machine.

If you are already impressed by genetic algorithm applications, prepare for the final strike. Finding solutions and designing everyday objects leads us to even more exciting areas of usage, particularly thrilling as these are still recognised as pure human domains. Welcome to **artificial creativity** and **artificial art**. Believe it or not, but the simple concept of following nature while operating on sequences of 0s and 1s results in creating systems able to generate jokes, paint drawings in the style of a selected artist (e.g. Picasso) or compose melodies inspired by a particular musician – you can already find all over the Internet Beethoven-like symphonies which he

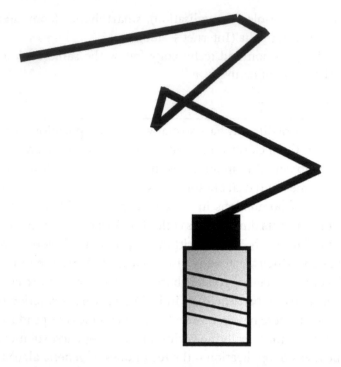

FIGURE 4.4 The quite unexpected shape of a small NASA antenna (the concept illustration).

never heard or played but which still would be recognised as his by critics and experts. Language is another area as well – in 2016, an artificial writer (i.e. a machine algorithm) wrote a novel that almost won… a national literary prize in Japan!

✎ **NOTES**

- Evolution is a natural process which drives the development and constant modification of all living things, making each next generation better prepared to live in a specific environment, e.g. run faster or hunt more effectively.
- Genetic algorithms squeeze the evolutionary process into milliseconds by manipulating sequences of 0s and 1s, called chromosome groups in populations.
- The algorithm works in cycles: selection, crossover and mutation.
- Natural selection is the mechanism used to select a partner for reproduction. Individuals stronger, taller or more likely to survive for some other reason are more often chosen as a partner by the opposite sex.

In IT, selection is based on the so-called fitness function which describes how close to an expected result the current chromosome is.

- In the world of nature, individuals group into pairs to give life to another generation. In AI, this stage of the algorithm is called cross-over: selected chromosomes (sequences of 0s and 1s) are divided into halves and the divided pieces are combined with others.
- Mutations are mistakes in chromosomes which may cause diseases (harmful mutations) but can also bring some benefits (like e.g. malaria-resistance genes in some indigenous sub-Saharan African inhabitants). In genetic algorithms, a mutation is a simple change in a randomly chosen position in chromosome.
- Genetic algorithms are applied in three main areas of usage: finding solutions (or answers to some nontrivial questions), designing objects and processes and artificial creativity and art.

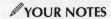 **YOUR NOTES**

CHAPTER 5

Monte Carlo Method: An Unexpected Benefit of Gambling

MONTE CARLO IS A district in Monaco, one of the smallest, most densely populated and richest independent countries in the world. This city-state is located on the beautiful French Riviera on the Mediterranean Sea and has the world's highest life expectancy, at nearly 90 years. But what Monte Carlo is most famous for are Formula 1 Grand Prix and gambling. Whenever you think about casinos and the stone faces of top poker players there are three places on the planet, each on a different continent, which they can call their home: Las Vegas in North America, Macau in Asia and Monte Carlo in Europe. The most famous casino, the Casino de Monte-Carlo, was opened over 150 years ago and there are no signs to suggest it could close any time soon. But what has this to do with a popular artificial intelligence method that shares the same name? They have one key common feature – they both win pots by drawing knowledge from the theory of probability.

Although strategies in a casino are quite an interesting topic too, we will keep our focus on computer science and artificial intelligence in particular. You may start wondering how things as unpredictable as dice throws, random card picks or roulette spins could be compared with professional methods used by engineers whose work by default is characterised by precision and efficiency. But the truth is that we all use the Monte Carlo method everyday, even if we have never heard its name before. Imagine you are shown a big box of Lego bricks and asked to state which colour of bricks are in the majority in the box. And here is the catch – you cannot touch the box or anything inside, you can see only the top surface, the first of many layers of the brick. Tough task, is it? Of course, we cannot be sure of the answer not knowing what is below, but we can still make a quite good guess. If most of the bricks we can see are yellow, we have no reason to doubt that the colour is the most frequent in the lower layers as well. If you find this example too abstract, let us look at two more, this time directly related to our everyday activities. The first refers to what we generally call quality assurance. How can you check that products leaving your manufacturing line are exactly as they were designed to be? In a chocolate factory, you are unable to verify (unpack and taste) each item produced – you would not have enough time and, more importantly, nobody would buy a bitten bar. So what do companies usually do? They pick random samples to check taste, consistency, texture, melting temperature and many other parameters. What is important is that the sample bars are picked randomly e.g. around 1 in 100 or 1,000 of the final products. If everything is fine there, then the company can assume similar quality for all bars in the series. The process is equal all the time so the idea seems quite correct. Let us study another case, taken just from your kitchen. You are preparing for your new friends a special soup that you feel especially proud of. The recipe is your secret, everything is almost done and now it is the moment to add spices. You have already added a little salt, pepper and a few bay leaves. But should you add a pinch of chilli powder too? What do you do? You simply stir up the soup (so that all the spices and flavours are more or less evenly spaced in the entire pot) and taste a single spoon of it. You assume the taste of the whole dish is the same. And so have the world's best cooks for generations. Random selection helps to create a culinary masterpiece. So why do not use the same technique for computer applications?

The idea of the Monte Carlo method applied within computers follows exactly the same scheme as the everyday examples described above: in order to draw general conclusions about an analysed object (which is too big to check it carefully piece by piece), a system analyses randomly selected samples. If the samples are picked well enough, the conclusion is quite close to the reality and we save a lot of time (and energy) compared to traditional, arduous detailed verification. Sounds awesome, right? But just what does it mean to pick a sample *well enough*? To meet this requirement there are two aspects that need to be considered: the number of samples to be chosen and the level of randomness used to pick them up. The first one is quite obvious – the more samples you have, the surer you may feel about the final results. If you analyse an object which is quite homogeneous (of the same kind) like tomato soup, tea with sugar and lemon juice, a ton of carbon, a huge roll of canvas, etc., then very few tries may be quite enough. After mixing his dish carefully, a good cook needs just half of single spoon to judge the taste of a 20 litre pot and feels comfortable to serve it to even the most demanding of critics. But if the object is more diverse and its parts differ from each other, like a complicated economic graph, a pool full of colourful balls, or the Amazon jungle (being home to thousands of plant species). In that case, more samples definitely need to be taken to realistically describe the whole. For example, with a single try only you may grab a yellow ball from the pool and a middle-sized tree from the rainforest, but the conclusions based on this strategy could not ever be said as even close to the truth: it would suggest that all of the balls in our pool are yellow and that the Amazon jungle consists of only middle-sized trees and only of one species – like a plantation of Christmas trees. So, the guidance on the number of samples is quite straightforward: unless you have a very homogeneous object which you are analysing, the more samples you take, the more precise out you get. Very often, the Monte Carlo method is used to find a numerical result and the question to ask is how exact the value is supposed to be. So, at the end of a day, it is the user's decision to make – are we fine with rounded numbers or do we need four digits after the decimal point? That of course depends on the application and just our whims – it does not really matter. What does matter is that you have just learned the first secret of the successful usage of the Monte Carlo method. The second one would again bring to our minds huge, bright casino halls and poker players carefully hiding their fortune-worthy cards – the words *chance* (or probability) and *random* will occur often in the following pages.

Let us start with a simple task. Choose a **random** number from 1 to 100. It could be 23 for example. Now, think for a while and pick again a random number of the same range. Are these two numbers truly random? They are definitely not. In the second try, you already know the first value and that implies the way you analyse the task and prepare for the second guess. Have you chosen 22 or 24 or have you thought of two numbers next to each other that could not be really treated as incidentally taken? We are much more likely to choose the second number from another fraction of the numerical range. In the famous Polish National lottery, you need to guess 6 numbers out of 49 to win the jackpot, so when I was buying a lottery ticket once with my friend I selected 1, 2, 3, 4, 5 and 6, the first successive options. You are unable to imagine the amount of criticism I heard then. Everyone said it was wasting money as such a result is almost impossible. But the chance (although less than 1 per 13 millions) is exactly the same for all the sequences. It is our brain which cheats us – it is simply not evolutionarily ready for random number generation, especially as this would mean the loss of other skills that we value much more: analytical thinking and learning based on experience, memory and feelings. That leads us to quite a paradox: the more primitive an animal is the more we can count on it in random selection (although it would never achieve perfection due to instinct, internal perceptions like hunger, environment influences like slight light changes). Let us momentarily go back to our task again. Have you considered at least for second to choose 23 again (so the two selected numbers are the same)? You may find it ridiculous but if you think about this carefully you will find it was never said that the numbers cannot be repeated. That is the most visible example of our limitations – some options are automatically cut from analysis. What about the first number picked? Unfortunately, this is also far from true randomness. You have chosen this value after conscious consideration and that makes it enough – you have made the choice based on your current feelings (e.g. *17 is my lucky number*), experience (e.g. *Last time I was asked for a number I gave 10 and the performer found it difficult*), physical state (e.g. *Feeling so tired. 1 to 100? 100 is fine.*), etc. It is never really random. Trust me. The same problem has faced computer engineers for decades. Surprised? Whenever software installed on your PC generates a "random" value (in games but also in much more serious applications like creating unique cryptographic keys) is never 100% true. Such values are called **pseudorandom numbers** and although they appear to be random on some level, they cannot be due to one crucial reason – they are generated by a software which is always a deterministic (not chaotic), step-by-step algorithm implementation. Pseudorandom number generators differ a lot, from simple math formulas to extremely advanced multi-machine systems used for security purposes by armies or secret services. It is worth noticing that the best generators start their calculations with a special data set named a **seed**, which is supposed to be as close to true randomness as possible (and so it is a combination of various factors like the user's mouse movements, pauses while typing on a keyboard, temperature of the computer components, memory usage changes and many more). It is never fully random; however, it helps to make the generated value look like that. And that is enough in most applications. So, are we able to find true randomness anywhere? The closest is probably equipment used by casinos, like dice, roulette wheels, and more, which, according with the law, need to pass thousands of independent randomness tests and be certified before they are allowed to be used.

The two above aspects are crucial for the precision of the Monte Carlo method: the number of samples taken (the more the better) and randomness (the closer to truly random picks we are the better). We have already talked about when we use or could use this technique every day: meal tasting (or better to say testing?), checking a tea's sweetness or guessing colours for a huge number of balls or Lego bricks. But I believe you are still waiting for something more unusual, a solution you maybe never considered before for a specific problem. So, here is a more challenging task to be completed: calculate the area of New Zealand using only a map with no scale on it. You are expected to say a number (of square kilometres) and all the information you are told is that each side of the map reflects a distance of 1,300 km. How to deal with that? You can easily calculate the area of the whole map, which is 1300 km × 1300 km which equals 1.69 million km². But what should we do next? Of course you can try to divide the area of New Zealand into dozens of small pieces and apply advanced geometric patterns. This could be quite a time-consuming challenge, especially if you imagine much more complicated maps of countries characterised by very irregular shapes and extremely long and ragged coastlines like Norway, Canada or Indonesia. However, there is a much easier way to complete the task – here is where the Monte Carlo method arrives to help. What we need to do is to throw (or select – it depends on the perspective but does not really matter) a specific number of random point on the whole square map. Let us try 20 points. Remember the randomness. The best idea is choose a point with your eyes closed or throw little paper balls on the map (Figure 5.1).

Then, count the number of points (or balls) that hit a region within the New Zealand borders. The method suggests that the ratio between that number and the number of tries (here, 20) reflects the relationship between the area of New Zealand and the area of the whole map. To find that ratio, simply divide the number of points in the country (here 3, see

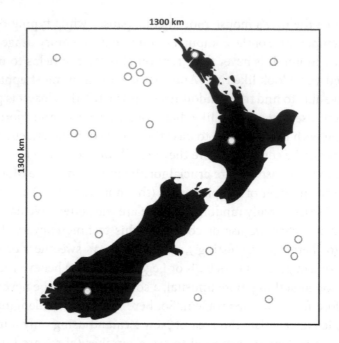

FIGURE 5.1   The map of New Zealand with 20 random points thrown on it.

my results in Figure 5.1) by the number of all points, so 3/20 = 0.15. We already know the area of the map, which is 1.69 million km². You may be quite surprised but we are just a single step away from the final result:

$$\text{New Zealand} = 0.15 \times 1.69 \, \text{million km}^2$$

$$\text{New Zealand} = 0.2535 \, \text{million km}^2 = 253\,500 \, \text{km}^2$$

We have thrown 20 randomly chosen points but despite this, the results is quite close to the reality – the area of New Zealand is exactly 268,021 km². Of course, our result is not the perfect one but it would quite enough for most applications – you would probably not fail an exam giving our value. And we can always improve the precision just by increasing the number of samples, in other words – the number of throws.

🚀 ROCKET STUFF: INDEPENDENT EVENTS

Probability theory may sound like a far-from-reality collection of theorems prepared by bright mathematicians in some of world top universities. But the truth is that this theory is the foundation of one of the most profitable industries ever – gambling. Have you ever wondered how it is possible for casinos to earn billions of dollars a year even though nobody is really able to predict the next roulette number drawn? It is all explained in an equation that seems so abstract to us at the first glance. The whole theory is based on the relationship between events being part of a sample space – a collection of all the possible situations that may occur in a specific context. Suppose you throw a standard six-sided dice. Then the sample space consists of six events. The chance (or probability) to find four dots on the top side is $1/6 = 0.17$. Probability in general is always a value from 0 to 1, where 0 refers to an impossible event (e.g. getting a 7 while throwing a six-sided dice), and 1 means an event that will surely happen (e.g. getting a result from 1 to 6 while throwing a six-sided dice). If we denote a probability of some event by p then the probability of a complementary (opposite) event equals $1 - p$. So the chance of getting all results except 4 while throwing a dice is $1 - 1/6 = 5/6 = 0.83$. The topic becomes a little bit more advanced when we consider some sequences of events. Here, one of the most important definition is event independence. Two events are said to be independent if the occurrence of the first one does not affect in any way the probability of the second one happening. Flipping a coin many times meets this require-ment: it does not matter how many attempts have already been made, the chance of flipping heads is always $½ = 0.5$. But be careful. The correct result depends on how you treat these events. If you wish to flip heads five times in a row, that means you make the events dependent and the probability is far from 0.5 – you need to combine (multiply) the component probabilities, i.e.

$$0.5 \times 0.5 \times 0.5 \times 0.5 \times 0.5 = 0.031.$$

If you would like to flip heads 25 times in a row you could try of course... but you have a much higher chance of winning the pot in a national lottery and that definitely sounds more useful I suppose.

Event independence is crucial in the famous **Monty Hall problem**. Have you ever seen *Let's Make a Deal®**, the television game show? Simplifying the rules, a participant chosen from the audience is asked to pick one of three doors (A, B or C) located on the stage. Behind one of them is hidden a brand-new car which the lucky participant could win as a prize, but behind the two remaining ones there is the *zonk*, a small mascot or even nothing. The chance of winning huge money is of course 1/3 but the game has just begun. After choosing the door (let us assume A is the choice), the host tries to make the show even more exciting. He

---

* The show was originally hosted by the Canadian-American producer Monte Halparin a.k.a. Monty Hall, which is where the mathematical problem got its name.

asks the technical staff to open one of the unselected doors, e.g. C, to let everyone gathered find that there is a zonk behind it. Now the probability of the participant winning is higher. And here comes the final question: "would you like to change your initial selection and pick B instead of A?" So what should the confused contestant do? Hold on to his or her initial decision? There are only two doors left, so we could think the chances are 50–50 and so the probability is 0.5 for each of them. And here comes the surprise. The probability for winning is doubled if the participant decides to change their choice! How is this possible? The answer is that the events are dependent, so opening the door denoted by C influences the further probabilities.

Let us look at this in more detail. At the beginning, the probability of winning while choosing A is 1/3 and this value never changes. That of course implies that the chance to find the prize behind door B or C (chance for B plus the chance for C) is equal to 1 – 1/3 = 2/3. And when C is opened, revealing a worthless zonk, the probability of winning with C rapidly drops to 0. However the winning chances for B or C could not change and are still 2/3 and that value now fully belongs to B. This result shows how powerful the theory of probability is. Imagine how many people could double their chances in the show knowing the basics. And remember the final conclusion – whatever you think of and no matter how hardly others try to make you keep your decision, it is sometimes really smart to change your mind!

HOW MUCH IS π?

Although mathematics clearly proves that the set of all numbers is infinite, there is one value which has been stimulating human imagination for centuries. The number is π (**pi**) – a constant representing the ratio of any circle's circumference to its diameter, usually approximated as 3.14159. Pi is an irrational number, meaning we are not able to represent it as a fraction (a quotient of two integers). What is more is that the digits not only do not construct repeating patterns but are sometimes considered as even an example of statistical randomness (although this has not yet been proven). For over 2,000 years, one of the most though-provoking geometric tasks was **squaring the circle**: to draw a square with exactly the same area as a given circle using only two simple tools: a compass and a straightedge. This problem was solved based on another feature of pi discovered in 1882: this number was found to be transcendental and thus the construction of such a square could finally be proven as impossible. This fact would have been a real blow to generations of mathematicians spending literally years on their attempts to complete the challenge. Although it may seem sad, there is also a lot of optimism there: first, due to their attempts, many other significant discoveries were made as "side effects", and secondly, it also shows what is most fascinating about science – you never know when the new big breakthrough will be announced. It may be in seven years, but it could also happen in just a few days. You can wake up one day and find out questions being asked for decades have been answered.

Despite many other interesting aspects, there is definitely one thing about pi that has been inspiring mathematicians most as it is a kind of never-ending race. The race for the next and the next digit in the approximation of pi. The first results appeared together with huge interest in science in ancient Egypt, Babylon and India: $22/7 = 3.1429$, and $\sqrt{10} = 3.1623$. But the real revolution in the approximation of pi came with Archimedes' method based on polygons – a time-consuming but exacting technique that was successfully applied for almost 1,000 years, consequently extending the space needed to write the known value of pi . The algorithm is simply beautiful, especially due to its simplicity. The idea is based on quite an interesting observation: if we look at plain polygons with more and more sides – starting with a triangle, then a square, pentagon, hexagon, heptagon, etc. – these shapes are continuously getting more and more similar to the shape of circle. Try and draw it yourself. So, what Archimedes suggested was to draw a circle and two polygons of the same type (e.g. two pentagons), one circumscribed (so the circle is inside and its line touches all of the pentagon sizes) and one inscribed (so the circle is outside and its line touches all of the pentagon tops). Now, we can calculate the area of both polygons in the traditional way (without necessarily knowing the value of pi) – the results give us the upper and lower bounds for the circle's area. The more sides the polygons have, the less space is left between them and so the area of the circle is more precise. Knowing the area we can find the approximated pi value using the well-known formula πr^2. Let us try and calculate the pi range for squares (polygons with four sides). If we assume, which does not affect the result, that the radius of the circle is 1, then the circumscribed ("outside") square side length is $1 + 1 = 2$, so its area is precisely $2 \times 2 = 4$. Now what is the area of the circle itself? The formula says πr^2, however r = 1 so the results is just π. Finally, we should calculate the area of the inscribed ("inside") square. When you look closely, you can realise we can say that the area of this square consists of four (right and

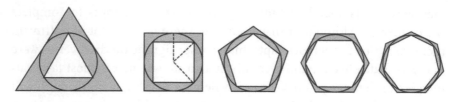

FIGURE 5.2 Archimedes' method: two polygons and a decreasing grey area between them (pi inaccuracy).

isosceles) triangles (see the dotted lines on Figure 5.2) so the result is the following:

$$4\times\left(\text{single triangle area}\right)=4\times\left(0.333\times1\times1\right)=4\times0.333=1.332.$$

Although there are just a few lines of the text we are quickly approaching the final conclusion. Just one, quite obvious, observation needs to be recalled: the circumscribed square is of course bigger than the circle (as it is outside) and the circle is surely bigger than the inscribed square. Thus, the relationship between their areas give us our very first approximation of pi: $1.332 < \pi < 4$. You are probably not excited about the precision we have achieved, but remember, it is always something more than a pure guess and, much more importantly, this result is for squares. If we consider polygons with more and more sides we will find better and better approximations. The space (between the upper and lower bound) for the guess, or error rate, or imprecision level – however we call it, this parameter (visualised by the grey colour in Figure 5.2) is constantly decreasing. Suffice to say, using this method, Archimedes proved that pi is surely more than 3.1408 and less than 3.1528. For centuries, mathematicians, both professional and amateur, were reached more and more digits using polygons with huge numbers of sides (and huge here means thousands). One of the hardest-working researchers was Dutch mathematician Ludolph van Ceulen, who spent a significant part of his life looking for further approximations of pi. Finally, he achieved 35 digits which were later engraved on his tombstone. As an expression of appreciation for his extraordinary contribution in this topic, pi is sometimes called the *Ludolphine number*.

Luckily, we do not need to spend our life finding a nice enough approximation of pi. As before, the Monte Carlo method arrives to help. As pi describes the relationship between a circle's circumference (but also the circle's area) and its radius, we can again start with basic geometrical calculations. We have already done some in the previous section while recalling Archimedes' technique, so don't be surprised to see clear similarities. Let us look at a circle and a circumscribed square and quickly calculate the areas of both (assuming, again, the radius equals 1):

$$\text{Circle's area} = \pi \times r^2 = \pi \times 1^2 = \pi$$

$$\text{Square's area} = (2 \times r)^2 = (2 \times 1)^2 = 4$$

So what we found if decide to divide the first area by the second one…

$$\frac{\text{Circle's area}}{\text{Square's area}} = \frac{\pi}{4}$$

So in other words:

$$\pi = 4 \times \frac{\text{Circle's area}}{\text{Square's area}}$$

If you are not a maths enthusiast, please to not leave the book aside. All of the formula's modifications are already complete. This book is for everyone and I am going to keep that promise. Can you see something interesting in the final formula? It shows how to simply calculate pi knowing the area of a circle and the circumscribed square. Of course, the more precisely these areas are determined, the more exact the pi approximation would be. Just how do we find these values (areas)? Sound familiar? We faced a similar problem earlier when we were stuck trying to calculate the answer to what was the land surface of New Zealand. This time we are going to follow the same concept. All because it is enough (although it is definitely not obvious) to make one smart observation: the truth is that to find π we do not have to actually know the mentioned areas. Instead we only need to find the ratio $\dfrac{\text{Circle's area}}{\text{Square's area}}$, which, believe it or not, is a much easier task. As in the case of New Zealand, we can prepare a square illustration with a circle inscribed in it and throw points in there. However this time I would like us to do it a little bit differently to increase the level of randomness even more. Let us draw lines dividing the original "map" into 100 small squares (ten rows by ten small squares), each with a number assigned – see Figure 5.3. Now forget about paper balls and use dice. Perfectly, if you are a fan of role play or advanced board games – then you probably have two dice – K10 (a dice with ten sides numbered 1 to 10) and K100 (a dice with

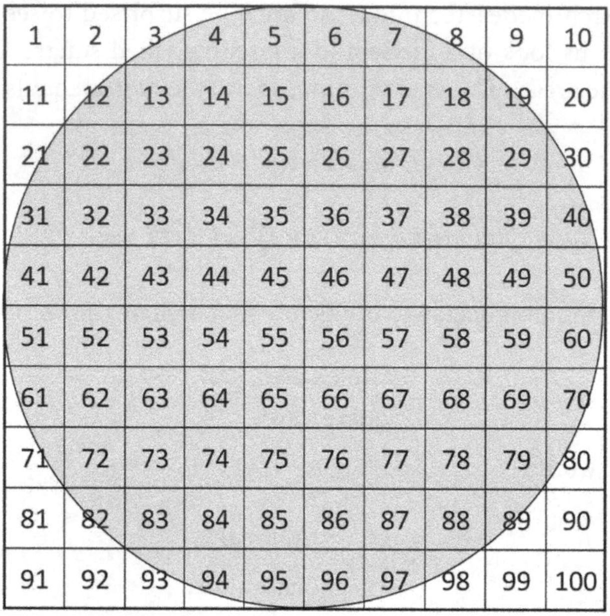

1	2	3	4	5	6	7	8	9	10
11	12	13	14	15	16	17	18	19	20
21	22	23	24	25	26	27	28	29	30
31	32	33	34	35	36	37	38	39	40
41	42	43	44	45	46	47	48	49	50
51	52	53	54	55	56	57	58	59	60
61	62	63	64	65	66	67	68	69	70
71	72	73	74	75	76	77	78	79	80
81	82	83	84	85	86	87	88	89	90
91	92	93	94	95	96	97	98	99	100

FIGURE 5.3   Circumscribed square divided into 100 equal pieces.

ten sides numbered in tens, i.e. 10, 20, …, 100). If you do not have these simple random devices, choose a digital one (available as mobile apps) or prepare an urn with 100 unique vouchers; whatever you prefer, make sure the method is as random as possible and that it is ensured that you can select the same value more than once (so in the case of an urn remember to return a selected ticket back to it before the next round). The next step, as you surely remember, is to decide the number of samples (tries) to be taken. The more samples, the more precise the approximation. Let us try with 20 samples. So the task for now is to pick 10 random numbers from 1 to 100 (dice, an urn, a mobile app – it is up to you) and write them down. Here are my results:

7	44
18	35
88	22
81	54 – *duplication but it's fine*
11	100
67	27
54	8
55	56
42	85
19	49

The second step to check whether each value falls inside the circle. If so write down "1" next to it, otherwise "0". In the case where a selected piece is partially inside and outside of the circle, e.g. the one denoted by "12", then mark it as 0.5. Do this carefully yourself and come back to check my list:

7	0.5	44	1
18	1	35	1
88	1	22	1
81	0	54	1
11	0	100	0
67	1	27	1
54	1	8	0.5
55	1	56	1
42	1	85	1
19	0.5	49	1

The next step is equally easy – just sum up your scores. In my case it is 15.5. And there is nothing left except to use our value with our special formula – we will treat the number of points inside the circle as a representation of its area, and the number of samples as an approximation of the circumscribed square. Nothing more, nothing less. Let us check how much pi is:

$$\pi = 4 \times \frac{\text{Circle's area}}{\text{Square's area}} = 4 \times \frac{15.5}{20} = 4 \times 0.775 = 3.10.$$

That means that our error rate is around 0.04, which is quite an extraordinary approximation of pi, given that it is based only on random dice throws, wouldn't you agree? Of course, if we divide the square into more pieces, e.g. 100 by 100 or 1000 by 1000, and take more samples, we can increase the precision without any difficulty – our calculation and technical skills are the only limitation. The same procedure can be also implemented with a computer. A simple programme could choose random points in the coordinate system and verify whether they fall inside a predefined circle (or whether the points [x,y] meet the circle equation: $x^2 + y^2 \leq$ radius). The more samples taken and the better the pseudorandom generator offered by the computer, the more of π's digits that will be revealed. Here are some approximations that I have achieved implementing a very trivial programme in the Java language:

$$10 \, \text{samples} \rightarrow \pi = 2.4 \left(\text{error rate} = 0.742 \right)$$

$$100 \, \text{samples} \rightarrow \pi = 3.28 \left(\text{error rate} = 0.138 \right)$$

$$1000 \, \text{samples} \rightarrow \pi = 3.108 \left(\text{error rate} = 0.034 \right)$$

$$10000 \, \text{samples} \rightarrow \pi = 3.1336 \left(\text{error rate} = 0.008 \right)$$

$$100000 \, \text{samples} \rightarrow \pi = 3.14244 \left(\text{error rate} = 0.0008 \right)$$

$$1000000 \, \text{samples} \rightarrow \pi = 3.14192 \left(\text{error rate} = 0.0003 \right)$$

So, as already said, the more samples we select, the more precise the result we are likely to get. Each time I describe the example, I feel a little bit amused as well – pure randomness let us find the value of one of the most famous numbers in mathematics. What took years for ancient Greeks we can get by throwing dice in a pub. The power of the theory of probability is much bigger than it seems to us at first glance.

However, calculating an area in a map or approximating pi are just the simplest applications of the Monte Carlo method. Exactly the same method is used by physicists to model liquids, gases or molecules, and medicine experts to analyse biological structures. But it is not used only in laboratories. The Monte Carlo method is used for assessing risk in business processes (simulating the consequences of various decision made by company heads to avoid failures) and preparing the perfect company portfolio while investing in the stock market – so even if you are a very serious and down-to-earth businessman starting out, do not try to have everything measured precisely. It is simply impossible – sometimes it is much better to have a dice in your pocket. People involved in production know the method well too – similar solutions are applied while designing integrated circuits or planning locations for wind farms or wireless network configurations. Even if you are none of the above, you see Monte Carlo results everyday – it is widely used in computer game AI engines and in light generation in 3D graphics (e.g. in virtual reality worlds).

Monte Carlo is one of the simplest AI techniques ever created and, at the same time, a method with an incredible number of applications. So, the next time you laugh at gamblers, think of the other side of the coin. The theory of probability has much more to offer than predicting our chances while playing roulette.

✎NOTES

- The Monte Carlo method is based on choosing random samples of the analysed object and drawing general conclusions based on them – exactly as we used to do in our everyday life (trying a few spoons of some soup to get the idea of the taste of the whole pot).
- The success of the method depends on two factors: the number of samples taken (the more the better) and the quality of their randomness.
- To achieve a high quality of randomness in computer programmes is not an easy task. No value generated by a computer is truly random, as it is based on a sequence of operations already defined. Thus, we usually call them pseudorandom numbers.
- A simple dice roll is "more random" than the most advanced systems ever implemented. That shows how far is still to go for a machine to be able to create a perfect simulation of reality.
- π (pi) = 3.141592... is a constant representing the ratio of any circle's circumference to its diameter. The number has inspired and fascinated people since antiquity.
- The Monte Carlo method is widely used in science, technology, business and virtual reality building.
- Randomness is much more exact than we imagine.

✎YOUR NOTES

Language Processing: Plato and Expert Systems

L ANGUAGE IS PROBABLY THE most important invention ever made by
humankind, much more significant than the proverbial wheel. The
appearance of language, as far back as around 200,000 years ago, allowed
more coordinated activities, like building constructions or efficient hunting. Still it is just the tip of the iceberg. Language and speech support
thought exchange which is key in philosophy and the basis for the scientific and engineering areas we know. If everyone had to work and discover
the world on his own, we would probably never have left the Stone Age.
Discussions permit brainstorming, while positive and negative feedback
drive the development of single people as well as whole groups and societies. Language is also a key element of the teaching–learning process.
Language empowers sociological and interpersonal interactions – being
able to name and express our feelings, we have a toolkit to build relationships with others. Moreover, saying what was good and what was not
acceptable became the foundations for the legal standards that lie at the
foundations of modern civilization. The invention of writing extended the
available means far more. It was a basis for creating documents, international treaties, money and trade. It also become a spark that started
history – just think of the old chronicles which give us unique insights
into the life of societies that passed thousands of years before our birth.

Writing initiated the rapid development of art too – people realised that they could leave their thoughts, beliefs, imagination and ideas to the next generations as physical objects. Such artefacts could potentially last far longer than their own lives. And what is no less motivating is ambition – the chance to have your name written with capital letters on the pages of global history has inspired thousands to think wider and to do something to be remembered forever.

The ability to communicate with others and the skill to use a language in particular is often highlighted as one of the main features that distinguish us from animals. On the other hand, it is one of talents we all have in common – irrespective of country of origin, education level (yes, even if someone is an illiterate he can still communicate by voice or through song), faith or style of life. We all learn this unique proficiency, unlike other skills, just by contact with other people – the more time we spend together the higher level we achieve; if we close ourselves to the world, we forget how to talk freely. We can even look at this topic a little philosophically: paradoxically, this most crucial ability of every human being is one you cannot train yourself in, you cannot buy, find or guess. Another person is the only way to have it. Speech is so obvious that we do not really appreciate it. You can notice its value clearly in the moment you lose it, although only partially – whenever you are lost in a foreign city where nobody speaks your mother tongue. Then you need to switch to some other popular language like English or Spanish and suddenly the communication is not as smooth and subconscious as before. Now, the true challenge starts if you land in some exotic place where only local dialects are known. Within a few minutes your situation reverts quickly – from a place of comfort you fall into a position where you need to work really hard to get food, find a toilet or transport and simply survive or leave. Such a case may seem abstract to you but just to give you a number to think of – there are more than 1,600 languages just in… India (that is not a misprint – sixteen hundred!). There are small and isolated villages where only a single person knows some other language to let the community to stay in touch with the outside world. Imagine you found yourself in such a place, add technological differences and multiply by cultural factors – and you might start to count only on your gestures (which should be used carefully too!). Yes, let us all value our language and speech skills. Still, the value and the influence on our life is just one way to look at this. On the other hand, these abilities are so common and somehow subconscious that we do not even realise how many complicated operations your mind needs to perform to keep the dialogue going. Our ears convert the sound (air vibrations) into syllables and words. Our minds construct the phrases and analyse them to find the meaning. The meaning must be matched with the context too – the same sentence may be interpreted differently depending on the situation. From an automation perspective, each of the above determines a true challenge to IT, being of such complexity that it constitutes a separate, fully fledged branch of computer science. As already mentioned a few times in this book: something simple and natural for a human is extremely difficult for a machine.

Natural language is the most fundamental tool to communicate with other people. It is a skill we have been developing as humankind for thousands of years and thus is so intuitive, quick and easy to use. So it is not surprising at all that we are working on similar interfaces to interact with machines. The success of small devices like Amazon®'s Alexa® or Google's Home® – which you can place near your fireplace and just ask for some important news or facts instead for searching for it over the Internet yourself – shows the trend is clear and expected. The next to go are other household goods, cars (some features are already there) and many more. Still, controlling machines with your voice, although spectacular, is just one of many applications in the domain of natural language processing (NLP). Important applications occur in the area of automated translation and the current solution make system output close to a professional human linguist. There are some tests on-going where various chapters of best-seller books are auto-translated into various languages and back to the original one... and in many case the output is of a quality close to the Pulitzer winners' own writings! So, if your job position is to translate documents, then be smart and look into some opportunities around as professional interpreter – live translation that touches also on cultural aspects and body language will surely stay as a human-only bastion for longer. NLP systems are also intensively used in text summarising. Imagine you have a 300-page book to read. That would normally take hours. However, you can use a system that squeezes it into just a few sheets while still containing the most important information. You remember the amount of time spent in the evening when preparing for literature classes at school? We would have probably all dreamt of such a device in the childhood. Similarly, you can find applications in information extraction, in other words: smartly looking for some important data in dozens of files. And it goes much further than a standard browser search. Advanced techniques not only allow us to find a place where a particular phrase is used (at it may appear in there in a totally different context). They rather scan the whole database, try to understand it and merge all the occurrences into one final, precise answer. Call it an artificial spy or a never-tired secretary – one thing is sure: these systems will soon change the ways in which we deal with documents... Generally speaking, with more and more advanced natural language processing mechanisms, human–computer interactions will evolve to higher and higher levels. The very first interactions were simply based on specific commends (keywords) being typed on a keyboard and getting some digital output on the console screen in return. That was the interface, and the boundary between our world and the machines' one was very clear and steady – simple commands, digital answers. Since that time, we jumped into graphical interfaces (desktops, icons, trash cans – all making us feel like a screen is actually our private office). A few years ago we started to use touch screens, which make the distance even shorter – a mouse, which was something in-between, is slowly disappearing. And today, all kinds of new interfaces are arriving: virtual reality (VR), gesture control (in console games and TV sets), mobile unlocking with just a smile. Fully natural voice interaction and linguistic conversation with machines is surely something ahead that will drastically blur the border between the worlds even further. Already we can sometimes hear an artificial telemarketer on the other side of the line, or online seller-bots helping you to put more into your virtual basket. In a few years, we will not be able to distinguish anymore whether it is a man or a machine talking with us.

Natural language processing is clearly connected with knowledge gathering. As humans, we usually share the information between each other using words and phrases (speech, letters, text messages, books, etc.). Of course, there are other means, like mathematical formulas, gestures or art (like emotional sharing via paintings, music or dance); still, for a significant majority, natural language is the key to communicating. It is no surprise that people are able to describe complex events or objects using their own words. And, on the other hand, hearing a description, one is able to imagine the things being discussed. That makes the modelling and mapping of the world somehow almost as natural and intuitive as the way we use speech and language. To define concepts, we use words and collect knowledge by matching and grouping such concepts. So it is not surprising that NLP techniques are started to become applied more and more in various applications that model the world around us. Such expert systems can be used in many situations, especially where an important fact comes from human informers rather than from IT databases. Just to give you an

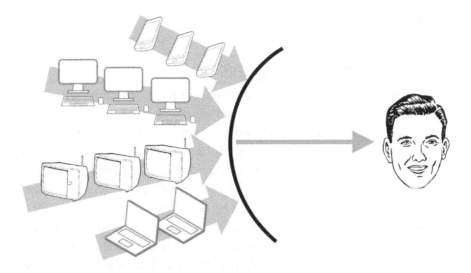

FIGURE 6.1   Intelligent information bypass.

example, we can imagine an system that collects messages from people gathering at, for example, a stadium (like during a soccer match or a big concert). These people could text a system to inform it about some risky or dangerous situation, suspicious behaviour or, in the other direction, asking for some detail. This idea was a foundation for the **POLINT 112-SMS** prototype that was designed for police forces to support mass events. The system is able to collect and process the information gathered and answer questions but also predict unwanted situations (e.g. fights, as two well-known hooligans approach each other) and raise it to the leading officers. One of the biggest challenges in any headquarters (in police squads, in hospital ER departments, in big IT offices) is dealing with the huge amount of information arriving from all directions (see Figure 6.1). Systems like the one described above, which understand the meaning and importance of such messages, are able to filter the crucial facts from the informational chaos, allowing us to ask for details, and highlight itself (without request) whenever something should be looked at. Such a concept can be named an **intelligent information bypass*** and it has started to change the way that we organise our services, work and private life. As mentioned earlier in this book, people today process more information daily than our ancestors in the Middle Ages did during their entire lives. And as there is more

---

* To find more various applications and the concept details please check my on-line available article (published together with Z. Vetulani ("Intelligent Information Bypass for More Efficient Emergency Management"): https://bit.ly/2NyRidG

and more information around, we may all soon need a private bypass like so we will not become overwhelmed by this mass of data.

SYNTAX: PLAYING WITH BRICKS

The foundation of every written communication is an alphabet – a set of letters that can be used to construct words. This set is usually somehow ordered so one can use it to number some items and sort words in an alphabetical order. Most languages spoken nowadays are based on the well-known Latin alphabet, which is derived from Greek letters (also still quite popular outside Greece especially in science like *gamma* (γ) radiation, tech, e.g. applications' *beta* (β) versions, military, e.g. US Army *Delta* (Δ) force, or marketing, e.g. *Omega* (Ω) watches). Still, it is worth knowing that the number of letters in an alphabet changes around the globe, from as little as 12 in the Rotokas alphabet, used by around 4,000 native people of Bougainville Island, located in the Pacific Ocean, up to Cambodia's Khmer alphabet, which has 74 letters. Based on the letters, a population can build words, which are like the bricks of interpersonal communication. Here again the diversity is extraordinary – some primal tribes use no more than a few hundred different concepts while the English language is believed to have the enormous number of 1 million words! What is interesting is quite obvious of course: no one applies them all, including Pulitzer and Nobel Prize winners. In fact, recent research shows that the average native speaker (born and growing up among other natives) knows around 20,000 words in the English language and after completing a higher university education, he or she is likely to double that number. Finally, it is recognised that on the everyday level, no more than 5,000 common words are used – if you imagine the full English dictionary as a book of 200 pages, an average person living in the UK rarely moves beyond the first page on daily basis.

A higher variety of words enriches a language, makes it more sublime, more melodical and more beautiful. Some psychologists even suggest that using a larger range of expressions to describe one's feelings helps one to live more fully and to avoid depression. It may sound quite strange but looking at this more closely, imagine a person describing his or her mood only as "good" or "bad". Then anything that what is not "good", automatically becomes "bad" and so intensifies the negative emotion. If the scale is granulated, then it is easier to find that it is maybe not as bad, as it could be much worse. That is why it is worth using more words to enrich our perception… Anyway, let us leave the psychotherapist's couch for now and jump back into the world of AI. As said earlier, a higher variety of vocabulary makes a language more interesting, but to make communication happen there is still one crucial element to be applied. And it does not really matter whether you use 5 or 50,000 nouns, you would not go any further without a cement that connects these bricks. The cement that connects word in a sentence is **syntax** – a set of rules and processes that formulates and modifies sentences. In the case of natural language processing, syntax is mainly related to grammar and identifying words as a particular parts of speech (e.g. noun, verb, adjective, adverb, etc.) and analysing parsing (which part is a subject and which is a predicate in a sentence). Now it is time for an example:

The man drives a new, red mustang.

Here you can find two nouns, but one of them is described by two adjectives. Both nouns are connected with a verb in the simple present. *Man* is the subject of the sentence, the most crucial part of it, while *mustang* completes the thought as a predicate. This is a proper phrase built according to English class grammar rules. All by the book. Unfortunately, the real world does not always follow the principles so directly: people use shortcuts, sentences are sometimes discontinued with some parts missing and finally people make mistakes. And they make them quite often. So, any system that is designed to support human–computer interaction needs to resolve syntax errors, which may not be trivial. And this especially refers to programmes which us the voice as an input, since it is easier for us to control what we write than what we say (that's also why it is always better to send an e-mail than to call if one is really angry). And it is not yet the end. In many languages, like in English, the order of words in a sentence influences the meaning, so if you change it the understanding may be totally opposite, e.g.

Cats eat mice. vs. *Mice eat cats.*

So, following the proper sequence of a sentence is another challenge for machines. What may be a tricky task is if someone uses slang or has a very specific way of talking like the famous Yoda, the Jedi:

Once you start down the dark path, forever will it dominate your destiny, <u>consume you it will</u>.*

---

* The quote is from Master Yoda in *Star Wars Episode V: The Empire Strikes Back* (1980).

Syntax brings many issues to be resolved for an advanced and efficient system. Still, this is just the beginning. We have the sentence about a guy driving a cabriolet. Just what does *driving* really mean? It seems obvious to us but not for a computer. Is *mustang* a type of horse? And who actually is *the man*, as certainly it is not myself? Without the force master controlling, we are left with few other technical methods to make an application understand what we say.

FROM WORDS TO READING BETWEEN THE LINES

Knowing the syntax structure of a sentence is crucial for the next step of any analysis. Without having it done properly, there is no further way to go. But, on the other hand, syntax alone won't tell us much about what the sentence actually means, why it was elaborated and what an interlocutor expects to happen after saying a particular phrase. So, in natural language processing syntax, the analysis of a text comes first, followed by another phase which is focused on understanding it. This is called **semantics** or lexical semantics (as a branch of linguistics) and is usually much more challenging as a task than the structural review done earlier. The reason can be found somewhere in between the richness of languages which have been developing for thousands of years, and the complexity of human perception based on the experience we collect during our entire lives. When a child is born, it does not understand a single word. While growing, though, the kid learns the world, new objects, new situations, new activities and quickly combines them with new words, phrases and expressions. As an adult, one understands most of the sentences they hear but never really all of them – it depends on the education taken, the profession specialised in, the experience gathered so far (for example, travels expand perspective and knowledge) or the community one lives within. Now, imagine a computer with no experience, no intuition and no knowledge about the world outside its motherboard. If you compare both cases you can easily see how complex and thought-provoking semantics analysis is.

Semantics is crucial for most natural language processing applications, like system control, automated cross-language translation, text summarisation or extraction, etc. In particular, human–computer interaction cannot be successful without it. If we recall classic movies like Stanley Kubrick's *2001: A Space Odyssey*, Ridley Scott's *Blade Runner* or Steven Spielberg's *A.I.*, we cannot be surprised that research on semantics been inspiring both scientists and artists for so many years. After all, the vision of a talking and understanding machine is a really a vision of an artificial person. A human-like entity created by humans.

Semantic analysis is not a trivial task. Suffice to say that some first attempts were based on directly converting or translating sentences as a sequence of words. So, each word had a precisely defined meaning. This only sounds easy. One of the fathers of philosophy and science in general, Aristotle, living in the 4th century BC, is often associated with the famous quote that the "whole is more than the sum of its parts". This refers more than perfectly to words and sentences as well. It is rarely possible to understand the meaning of a sentence just by knowing the meaning of the words it contains. A single word usually has more than one meaning, and there can also be groups of words sitting together, called **collocations**, that affect proper interpretation. As an example, look at the following four simple phrases:

hot chilli pepper
hot girl
hot kitchen stove
hot news

Although the word *hot* occurs in each of the phrase, each instance has a totally different value (meaning). The noun that follows the adjective in each phrase is clearly crucial in determining how to understand the word and the whole term. Of course, it is not always so straightforward; as an example, if we simply say *hot dish* without further context, we might be unsure whether *hot* refers to the temperature or the spice level. This shows the cruel truth: any direct, word by word, translation is naturally burdened with a risk of error. And as long as it is like the above example, the consequence can be a funny output, or an awkward situation in the worst-case scenario. However, if we imagine global peace meetings or huge international contracts, some misunderstandings may lead to months-long diplomatic crises or significant losses of revenue. Words are tools. Sometimes more powerful than guns. Precision here may save money, relations and... lives.

There are various techniques for semantic analysis and there are more and more new ones appearing on the horizon. We will not go through many of them – this is not because they are not interesting, more so that the idea of this book is not to be an encyclopaedia but rather a guide that shows ideas and concepts, and so inspires for further in-depth learning. At least, I hope it will work like that...

One quick and easy method of semantic analysis is looking at **keywords**, as simple as it sounds. You may think that it stands in the opposition to all that was said earlier – single words may not help to understand the whole sentence – but this depends on the purpose. Of course, complete analysis of a sentence based on some specific words in it will not be possible. Still, one can use it successfully to support human–computer interaction. A quick but, of course, not a smooth solution. How does it work? Simple: check if the sentence said by a user contains any or a required number of particular words. Then, suggest an interpretation and validate it with the user. This surely extends the communication by extra questions but, on the other hand, ensures confirmation of the message's meaning. Here is an example:

User: ... ... *scan* ... ... *viruses* ...
System: *Would you like to scan your computer to look for viruses and other security vulnerabilities?*
User: *Yes.*
System: *Roger that.*

So when seeing (hearing) *scan* and *virus* said together within a single sentence, the system guesses that the user's request relates to security scanning. Sure, it could mean something else, but that particular variant is of quite high probability – a user would rather not discuss his health condition with a random machine, telling it about viruses and x-rays in a medical context. If you think about asking the next question, the answer is *yes* – although this may not be so obvious at the first thought, mathematical statistics empowers a lot of varied solutions and methods across the area of natural language processing. The keywords technique is also the main point of one popular urban legend. The story says that the US National Security Agency has the means and technology to monitor any messages sent over the Internet. So if you send an e-mail containing these two words together: *president* and *bomb*, then it will be delivered to the recipients a few minutes later just to give NSA's systems time to crawl through the e-mail and analyse it more carefully. Whenever the legend is true* or false, it shows a very nice application of the keyword solution

---

* I assume it is an urban legend. Just in case it is not, then this book may be crawled as well, so here is my statement: *The example given in this book is based on random stories from the Internet. I never met anyone from the NSA... However, it would surely be a pleasure to have afternoon tea with your agency executives. Happy to arrange a date, guys!*

– nobody is able to interpret all documents quickly, but you can still quite easily filter out ones that should be considered in more detail. It is a kind of early selection (of course not a perfect one) to identify important data among petabytes of useless information.

The above methods are quick and simple, but to fully understand a received message, the system must understand the real-world relationships between the objects we talk about. And to do so, today's scientists found inspiration in discussions started by ancient philosophers over two millennia ago. So here is a pinch of history. It was probably Plato, living in the 5th century BC, who first started to more widely share his thoughts on the origin and structure of the objects which the world is built of. To describe his famous concept, he used an allegory nowadays called Plato's Cave: we are all like a prisoners kept in a tiny cave, tied up in such a way that we cannot see the entrance or the bonfire outside. All we see are the shadows on the wall in front of us. But these are not real things, only the reflections of real objects existing outside of our perception. Similarly, humans, here and now, cannot really see true things (ideas) as they are. We only see the shadows of the real world. If we think about this for a while, we discover that the *Matrix* movies are not so far from this ancient concept. The discussion on the structure of the world has been continuing since the time of Plato and will probably never end. Plato created the whole field of philosophy called **ontology**, the study of beings and types of beings, becoming, existence and reality.

Aristotle, Plato's most famous student, continued and extended the theories of his master teacher. One of his great ideas was to propose the so-called **genus–differentia definition** of any existing object. So, any definition contains two parts:

- Genus – what family does the object belong to?

- Differentia – what makes it different from the rest of its family?

So for example a square is a rectangle (genus) with sides of the same length (differentia – that feature distinguishes a square from other rectangles). The same technique is used in biology, so for example a mammal is defined as a vertebrate distinguished by hair, middle ear bones and mammary glands. If you look at any definition on Wikipedia, you can find similar way of describing topics. And exactly the same technique in used in modern AI solutions to build a kind of map of concepts to allow a machine to understand relationships between them. Yes, the 2,000-year-old definition is applied to world-changing technologies in the epoch of DNA modification and space travel… The fundamental piece of this artificial map is the **synset**, also known as the synonym ring, which is a set of synonyms, the linguistic elements that from a semantic point of view are equivalent. So, in other words, a group of phrases in a sentence which we can exchange in any sentence without changing the overall meaning. So for example we can consider the following a synset:

(car, auto, automobile, motorcar)

and so, in the sentence *The man was driving a new, red car* the word *car* can be replaced by any other word from the synset while the sentence meaning remains unchanged. The collection of synsets that presents the dependencies between them is called an ontology. One of the biggest (containing over 100,000 concepts) was created by the researchers at Princeton University*. There are various types of relationship between synsets (concepts). The most important one (that follows exactly Aristotle's definition) is called the **hyponymy** relationship (from the Greek 'hypo': *under*) that connect two synsets, where the first one is a special variation of the second one. For example:

car is a hyponym of *vehicle* (its hypernym, from the Greek
  'hyper': *over*)
walk is a hyponym of *move*
blue is a hyponym of *colour*

Another popular dependency is **meronymy**, which describes the relationship between the part and the whole, e.g.

wheel is a meronym of *car*
step is a meronym of *walk*

Ontologies like these become crucial in complex systems, especially expert ones that are designed to support human decision-makers. So, if such a system finds information about a knife or revolver, then, thanks to built-in wordnet structures, it can quickly identify it as a hyponym of a weapon or dangerous object. Understanding this, the application can immediately notify its supervisor so further action can be taken.

---

* Play with the Princeton WordNet* free at http://wordnetweb.princeton.edu/perl/webwn.

It was around 15 years ago when I visited the United States for the first time. I was really thrilled and excited to cross the Atlantic Ocean and to present my research at the Florida Artificial Intelligence Society conference. I stayed in a small motel in the surroundings of Fort Myers. So one evening, I asked the staff about the nearest shopping mall. I heard: *left-hand side direction, 10 minutes*. So I left the motel and started to walk. After a quarter of an hour I found myself in the middle of dark, empty road with no sidewalk for pedestrians feeling a little bit stressed as finding the scenery similar to that from *The Hitcher* with Rutger Hauer. There were some lights among darkness ahead and it took me a longer while to reach them. It was a gas station. I entered and repeated my question:

- Follow the road, ten more minutes – I heard.

- Are you sure? A ten-minute walk? – I asked to double-confirm.

- Walk?! – the man stopped his current work, looking at me with total surprise on his face.

The story is quite funny from the perspective of time and shows the most fundamental aspect in the semantic analysis, which is the **context** of a message. I am used to walking a lot, which I find the most pleasant and comfortable way of travelling, while the United States is the world capital of automobility, influenced by the Henry Ford revolution. You will not find many pathways there outside the centres of big cities, and having a car is more important than having a place to sleep at night. It is the mix of tradition, culture and fashion that has constituted that way of thinking for years. But context may also depend on the current situation in which the analysed phrase is said. Some examples are sentences about spatial relations, such as those mentioned while discussing the aquarium metaphor in Chapter 2. If you think of a simple sentence like *The object is quite far, on the left*, even here we can find difficulties in interpretation without any extra context. Where is *left*? Does it refer to the sender's or the recipient's perspective? How far actually is *quite far*? Do we speak of the scale of inches or maybe miles?

The role of context is also another concept. **Sub-languages** are specific combinations of words, styles and interpretations used by particular groups of people. In fact, every group of professionals uses their own sub-language. If we imagine ourselves taking part in a doctoral body in a hospital that discusses some difficult cases, then we would probably understand little to nothing of what they say. If you sit in a coastal tavern with sailors who spend most of their lives at the sea, you may find it difficult to follow their thoughts about the repairs needed on their boats. That is why AI systems are usually designed for a particular group – the diversity of occupations, contexts and knowledge levels is just too huge to make semantic analysis easy, even for human beings. If you have the sentence *Bulls are coming!* the meaning is absolutely different if these three words are said during a hard day on a farm or in a dark underground area that has no good fame in a city…

An important area of semantic research that is becoming more and more popular is **sentiment analysis**, which focuses on the interpretation of feelings and the intention of the author of a message. It is especially interesting from the point of view of the growth in online marketing as well as the influence that online communities are starting to have on reality. For companies and producers but also politicians it is becoming crucial to go quickly and automatically through thousands of comments below a post or a product page and understand the passions, needs and desires included in them. Of course, as you can easily imagine, this is a very challenging task. For example, some real emotions may be hidden behind the wall of irony. The sentence *This movie totally changed my perception of cinematography* may be an extremely positive review or just a smart way of saying an exactly opposite statement.

DATING ROBOTS

I remember the story of an old American Indian who was the last member of his tribe. He lived alone on a huge area of the tribe's reservation with a dog as his only companion. The sad thing was that at the same time he was the last person to speak the tribe's language and, as it knew no other, the only recipient understanding his message was the dog, which was actually quite well-trained and knew many tricks. When the man died, the dog was the last living inhabitant of the area and the last representative to understand the forgotten language. Paradoxically, there was no one able to command him or ask for a trick…

Although it may be quite surprising that some languages get endangered exactly as animal species, quite often the reasons are pretty similar: globalisation, transport development and technological expansion. Although there are still around 7,000 languages spoken around the globe, a few of them simply go extinct every year. Here are just a few examples: Fanny Cochrane Smith, who died in 1905, was the last one to speak the language of Tasmania (south of Australia). Her Aboriginal songs, preserved on wax cylinders, are the only sounds remaining of the language and are now a part of the UNESCO World Heritage register. Ned Maddrell, a fisherman living on the Isle of Man (between Ireland and the UK) was the last surviving native speaker of the Manx language and, besides being a kind of local celebrity, he also supported activities to revive the language until his death in 1974. Doris McLemore lived in Oklahoma (USA) and before she died in 2016, she spent almost 50 years teaching and preserving the Wichita language of local indigenous tribes of which she was the last fluent speaker. These three people are far from each other in both space and time, living on different continents and in different times. But they have one thing in common – the wish and strong motivation to keep their languages alive for future generations.

I said all the above to present an aspect we may not be aware of while living regular life – people identify themselves with the language they speak and consider it an extremely precious artefact. Communication is one of the most fundamental needs of any human being. We may not see it on a daily basis but the exchanging of messages is especially important for psychological stability and comfort. That is why even in the hardest prisons, separation from other prisoners in single cell is such an effective punishment for disobedience. Similarly, cutting someone off from the external world by blocking regular letter exchange or Internet access is also a popular form of pressure.

It is all because people by nature look for communities or groups to stay in contact with. This is a deeply encoded need inherited from our ancestors, who lived in an old, dangerous world. Having no technology to support them, being part of a group was crucial to staying alive. Otherwise, one would quickly be eaten by some predator, start to starve (smaller area able to search for berries and, let us be honest, you won't hunt a mammoth on your own) or die in suffering if hurt and left with no one to look after them. Communication, especially voice communication, helps us to fulfil this internal need. Everyone needs someone to talk to. People also grow tired of visual screens (TVs, laptops, tablets, mobiles) and something other than visual interfaces which allows them to control machines may bring a lot of freshness and comfort.

But that is not all. Today's society is changing, with more and more relationships moving from reality to the virtual world: social media replaces live meetings, shared documents are edited online instead of working together within one room, take-away food ordered via the Internet is becoming more popular than visiting a restaurant. These are the consequences of course of the growth of technology and the increased speed of every-day life. However, verbal communication is important for people's health and so systems responsible for natural language processing will surely become more and more popular. Currently, the biggest challenge is that spontaneous and off-topic chats are difficult to follow for artificial bots. That is why the famous Turing test still remains unbeaten after so many decades. Still, recent NLP techniques draw on various quickly developing AI solutions (including deep learning), so the breakthrough may be quite close. And once it happens, it will surely massively change the world we live in. Smooth and more natural conversation will not only increase our trust of computers (as they are able to work well) but also, and this is where the revolution is coming, make the machines less distinguishable from human beings.

So imagine a tool that speaks like a real human. Then it is not just a tool anymore. Natural language processing technologies inject a human factor into such a machine. This means that suddenly you are not only able to ask the computer for a favour, an action or information but also you can simply chat and enjoy spending time with. The relationship modifies while the conversation is kept active. The tool slowly changes into an office clerk, then a close assistant, a workspace colleague and finally into a partner. And if at some stage one can find the system to be his or her friend... or a lover.

✎ **NOTES**

- Language (to elaborate our thoughts) and writing (to preserve such a message) are the two most important inventions we have ever made as humans.
- Natural language processing is a crucial part of artificial intelligence as it brings a new interface to human–computer communication while at the same time greatly changing our relationships with machines.
- There are dozens of applications for NLP, including automated cross-language translation, information extraction or search and many more.
- Syntax analysis reviews a sentence structure, while semantic analysis allows its interpretation.
- Aristotle proposed defining any object in two parts: the family it belongs to, and the features that differentiate it from other entities in that family. This definition recently became the basis for the modern wordnet concept that helps machines to *understand* relationships in the real world.
- Context is crucial for any interpretation to be correct.

- The author of this book dreams of a **new, red mustang**.

(Is this a true statement and real wish or just a joke? Or maybe a suggestion for an action to be taken by happy readers? Or simply form of an auto-irony?)

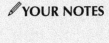

YOUR NOTES

CHAPTER 7

A Future with Artificial Intelligence

I REMEMBER WELL THE VERY first time I realised there was a concept like artificial intelligence, I heard about robots doing more than mixing ingredients for a cake to be baked, and about a future which might not be as bright as I had used to think. I was an eight-year-old boy at the time and I was spending the whole school year in Ciechocinek, a small health resort in the middle of Poland, famous for its saline springs and for one of the biggest graduation towers in the world. I was there to improve my health and respiratory system to stop sickness occurring so often. That evening, I was sitting in a small room with a TV set together with other, mostly older boys. The carer looked at me for a while, maybe wondering whether I was old enough, and then put the VHS cassette into the device. The movie was *Terminator 2: Judgement Day*, and I watched it with my eyes and mouth open wide. I cannot say if the film changed my life, but it sure influenced a young mind – I started to ask more abstract rather than everyday questions, and much more often looked at the sky than beneath my feet. I still like to come back to that movie and watch again some of its most remarkable scenes (like the liquid nitrogen tank destruction!) but as I get older I think of this story as less and less likely to really happen. Does it mean I do not believe in the era of machines? Certainly not, I am even more sure about this, just a little bit more optimistic. Personally, I believe in the unstoppable development of science, and whenever I hear the word *impossible* referring to the future of technology, I just smile and usually negate.

The problem with imagining the future is that we are strongly limited by our present way of thinking, limited by everyday concepts, habits and lifestyle. During my childhood, I had a friend who lived in the village. He was a smart boy but had no experience with technology. I remember when we were 13 or 14, he said he could not imagine talking through a phone as he had never done it (the next day I took him to the town, we bought a phone card and he has made his first phone call from the telephone booth – he was really excited). So, the reason why we hear so many concepts of the future and the word *impossible* is repeated so often is that we live today, in the current world. It is not only that we do not know yet all the answers but it is even more – we do not know yet most of the questions that can be asked. There is a famous statement, traditionally attributed to Lord Kelvin, said to his colleagues from the British Association for the Advancement of Science in 1900: *There is nothing new to be discovered in physics now. All that remains is more and more precise measurement.* Just five years later, Einstein announced his theories of relativity that overturned the foundations of physics. Lord Kelvin was not a random person – he was famous scientist and engineer who, among other achievements, formulated the laws of today's thermodynamics and determined the exact value of absolute zero (so the temperature unit Kelvin was named to honour him), so it not surprising that many people today do not believe that computers may dominate our word any time soon.

During discussions of the future of artificial intelligence, we usually consider the moment in time when it is likely to achieve a human level of intelligence and consciousness, allowing the system not only to answer questions asked by people but also work creatively, proposing new solutions and formulating concepts even when not directly asked. Such a future moment we call the **singularity** (or technological singularity). If we, as a civilisation, get to this point the consequences are really difficult to predict. First of all, a conscious machine would be able to continuously improve itself, creating newer and better versions of the system. Being connected to a global network and thus all the human knowledge stored therein, this kind of process would surely evolve from linear to much faster exponential growth (see Figure 7.1). In other words, such systems would be able to expand themselves much quicker than with the help of human engineers. The evolution of science and technology took us thousands of years from the invention of the wheel to a trip to the moon. Exponential growth could reduce this time to single years. We could expected solutions to some of the current most crucial civilisation problems (like deadly

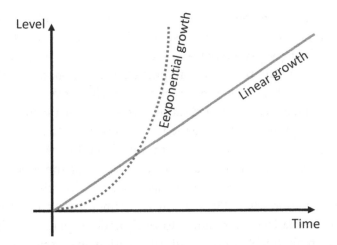

FIGURE 7.1   The difference between linear and exponential growth.

diseases) within a few months. One thing is certain – the singularity will be our last and biggest invention ever – all further discoveries will be made by machines, which will become faster and smarter researchers than any human could be. Is the singularity our future? I am sure about this. When will it happen? Some experts say in 70 years, some (including the ones who work on this everyday) predicts it may appear within the next five years. My guess – I would pick some value in the middle. I believe and truly hope to see the most extraordinary change in the history of mankind myself.

All that may sound abstract and *impossible*, but remember the phone booth story. Just 20 years ago, the concept of the Internet was out of mind – difficult to understand and discuss. The president of IBM® in the 1950s, Thomas Watson, was said to have once stated, *I think there is a world market for maybe five computers.* So do not close yourself within the current barriers – look wider, dream and look at the sky.

Nobody is sure about the future in front of us, but there are some pieces of it that I think we can predict with a higher chance already. To understand it better, let's start with the idea of a black box.

BLACK BOX

As I mentioned earlier in this book, recent research results suggest that people nowadays perceive more information every single day than our ancestors did during their entire lives in the Middle Ages. In addition, the science and technology develop with such speed that we never experienced before in the history of our civilisation. It is enough to realise that ancient philosophers like Aristotle contributed to almost all areas of science known in those days, starting with philosophy, through logic, kinetics, optics, astronomy, biology, psychology and more. Their research and discoveries influenced the future that we are all part of. Today, nobody is able to find himself comfortable in all branches of science. The extremely high level of knowledge and expertise needed caused even mathematicians to split into many focused groups and create theories so advanced that they are rarely understood well outside of this community (even by other professors of mathematics). It is said that when Andrew Wiles first announced his 200-page proof of the famous Fermat's Last Theorem (which had reminded unproven for more than 350 years at that time) it took months for the world's mathematician community to verify and accept it. A single human person is simply unable to cover all information, even that related to his or her specialisation – the world is changing too fast – some estimates say there are approximately 2.5 million new scientific papers being published every single year. And this number constantly increases! It is not surprising then that we simply take most things for granted, something obvious, not really having the time and capacity to check or understand it. We use dozens of tools and devices (probably you use at least a few just while reading this sentence) while not actually knowing what is inside. All we know is the external interface we use: buttons, touchscreens, switches. The rest (being in fact the bigger part of the object) is outside of our perception and interest, somehow hidden or invisible to our minds. We call these phenomena **black boxes**. We learned earlier in this book that IT people often use this concept to note that for some computer program or hardware device they are not interested in the details of its internal algorithms or the technical solutions of the way it is constructed. But the truth is that black boxes are everywhere around us, far from the IT applications and the latest technological gadgets. Nowadays, we do not know precisely how the banking system works (and where exactly our money is when it is already paid to our accounts), what our regular food products are made of (and how), what really happens when we call for a pizza (we just eat not wondering who baked it, where and how they did so and how it was delivered). Yes, we look at many things around us simply assuming they are all designed to fulfil our requirements, answer our questions or act in response to our requests. The 21st century is the world of services (with all the possible meanings of that noun) we really know nothing about. We enter the input and get back the output, not spending a second to think about what makes the first one to convert into the second one (Figure 7.2).

FIGURE 7.2 The black box.

What is important to notice we can see the clear trend in this topic. Karl Popper, recognised as one of the 20th century's greatest philosophers of science, identified three reasons of which at least one needs to be true to make any new theory (in general) replace an old one: the new theory is either more general, more precise or simpler. This makes the theory accepted as "better" by scientists. Exactly the same set of features makes other things more popular, more useful or more frequently bought. Believe it or not, but these three ways of evolution drives world changes, from IT to agriculture. It also causes a slow and constant enlargement of black boxes across various branches of science and technology. Let us think of the history of modern computer science. It all started with the electro-mechanical calculating devices that were built during the Second World War – one of the most famous is probably the Bombe constructed by Alan Turing and his colleagues which allowed them to decode the Nazi's secret Enigma machine messages and thereby increased the chances of the Allies winning the biggest conflict our civilisation has ever witnessed. The creators of such devices knew exactly how they worked – they were able to switch or modify specific parts of the machines to make it work faster or to fix calculating mistakes. Still, as computers started to attain the shape they have today, programmers began to work on higher and higher levels – at the beginning, they operated with binary codes, then from simple to more and more complex commands, finally working with more advanced frameworks nowadays. The truth is that these days, most software programmers work far from simple math-like operations and even further from the hardware itself. Using huge frameworks that provide many features ad-hoc, everything below this level becomes a black box. Software programmers from mathematicians, through inventors, to scientists and craftsman, finally become technically advanced users. They follow good practices, outside suggestions and precise standards. There is less and less space for free thought and trying unexpected solutions – all due to the global trends focused on time and delivery. IT people are specialised into narrow topics and their black box is bigger than ever before. Most IT techniques and solutions are taken for granted. That is one of the reasons why we suddenly heard recently of the vulnerability (security hole) affecting most of the world's microprocessors – the low level hardware is usually ignored by IT community, being treated as a black box which was and always will be there…

A black box is a common phenomenon and the area covered by it is extending. And what is especially interesting around the artificial intelligence methods already used is that the black box effect is visible there much more directly than in other branches of science. Why? Imagine an application based on artificial neural networks. As we already know, the learning process is mostly based on presenting learning samples to the network and providing an expected answer at the same time. When the learning set (collection) is big enough (and well-prepared enough) the network starts to recognise the patterns hidden in the samples and soon becomes able to answer correctly questions never encountered before. But what do the programmers actually do if it makes some specific mistakes? Do they review its structure and manually modify the weights, or add or remove particular neurons? Not at all! It would surely be too time-consuming to identify "the location of the error" in a network of hundreds of layers and thousands of single neurons. It is much easier to prepare one more example to show to the network how to answer correctly to the problematic question. If you think of this for a while you can find a quite clear analogy. That is exactly what we do when teaching others. Nobody really modifies anything manually in somebody else's head (ok, almost nobody – there is some research on neural surgery that may help in some physiological disabilities). It is surely much easier to explain a difficult example to a student one more time and to try to somehow inject the correct answer into a brain… although some teachers might not agree with that statement. So the interesting observation is that from that point of view we already look and interact with the AI (not being anywhere close to strong AI yet!) much more as we do with people than with other devices. We do not even know the thoughts of people who are our closest friends. Similarly, the AI becomes a huge, big black box.

When reading the above you may think I criticise today's world, being a population of consumers not interested in the point from which the products, ideas or services they receive come from. But I am far from disapproving of civilisation and saying we should all try to know and understand everything. In my opinion, such an attempt would make science and technology develop much slower or even fully stop it. So do not feel disappointed. A black box is a natural way our brain works to filter the important facts and let us (by filtering) deal with enormous amounts of information. We cannot keeping digging into details. We need to assume some foundations to build a skyscraper. However, what is important to just remember is that the black box is always there. Simple awareness of that fact is often enough to see much more than others.

POSTPONE YOUR UNEMPLOYMENT

Since the very first AI program was run and successfully applied. it became clear that sooner or later, artificial intelligence would change the future of our civilisation. It is again nothing new that technological progress changes the way we work and live. When fully automated production lines started to become more and more widely used, we realised it would directly influence the world's labour market. One computer-controlled line replaced dozens of manual workers. Huge factories previously employing hundreds of people are now working effectively with just a few members of technical staff. And as artificial intelligence becomes more and more popular, such a trend will only speed up. Scientists, economists and analysts agree on this aspect of future prediction (something which does not happen often): most of today's professions will sooner or later simply disappear, fully replaced by machines, and many of them will become extinct within the next couple of years. If you think of automated, GPS-based navigation systems or free-to-use online translators across an astonishing number of world languages, you can quickly notice for yourself that human occupations traditionally related with these activities (and usually requiring years of higher education) might not survive long. And there is not much we can do about this – it is all about simple and brute economical profit–loss accounts being the bloodstream of each existing company. If you can have something done for free by a machine which is never tired, never on sick leave, never complains, and can work effectively seven days a week 24 hours a day, you are definitely less likely to hire a human person. The scale of this trend is already visible in some industries and has started to become a topic in international discussions. High levels of unemployment are surely one of the factors that every society tries to avoid as it disturbs the cross-generation financial balance that affects medical and retirement benefits. People unable to work (because of machines taking their place) cannot achieve the standard of living they are used to and save money for old age. Besides all the other negative effects, it indirectly increases the chances for unrest and higher level of criminality. That is why there are already plans to include robots in human tax systems. Companies using machines instead of robots might be obligated to pay dues (as in the case of human employees) to keep national accounts within safe limits. Solutions on the government level are surely important but is there something we can do from our side to postpone our unemployment and stay secure in the labour market?

Someone once said that the key to career success in any company is to make sure you are irreplaceable in your everyday duties. The more people perform your usual job activities, the less sure you are about your future in there. It is not surprising at all that the same golden rule should be applied whenever we feel uncomfortable due to recent rapid AI development ments in our branch of the market. So, the first way to do it is to be a high-level expert in your domain. Even if machines are to replace your profession, it means job reduction in the first stage rather than firing all employees. Even in the further future, there will still be a place for the best experts – at the end of a day someone needs to control the AI, perform its periodical testing, monitor its behaviour and be the teacher or mentor to the system. These kinds of activities are surely the ones that are expected to be in value. So whatever you do – do it at your best. It is also worth mentioning that the changes in front of us will not only reduce and eliminate some professions but at the same time may restore to the world the ones that have remained unnoticed and underestimated in previous decades, described as not useful or too abstract. The voice of anthropologists, who currently investigate distant Amazon tribes and excavations, may soon be looked for by top experts and politicians to find and understand how AI is modifying our lifestyles and social skills and what risks are hidden in this modification. Behaviourists, nowadays sometimes treated as eccentrics and pet physiologists, will surely become well-paid and desired employees – the more independent machines become, the more crucial it is to understand their goals and behaviours (especially in the case of strong AI). Ethicists discussing the aspects of good and evil in university cathedrals will be asked to prepare the rules that computers should follow, e.g. to teach it how to choose wisely (and honestly) between two requests from two different users which are in conflict. Finally, philosophers, usually ignored by tech people these days, might be the only ones to be open-minded enough to talk about AI objectively and to try to predict the future of our civilisation. This list could be extended even more...

On the other hand, what is no less important is to distinguish ourselves from the machines. To protect your position, make it special and prove your actions would be difficult to replace by automated algorithms. The more repetitive the work is, the bigger the chance some of its parts will be computer-controlled soon. If you do not want to be replaced by an automaton, do not behave like one – do not act automatically, avoid work according to scripts and checklists, focus more on your experience and intuition (the thing that might never become a feature of a machine). Instead of following stereotypes and good practices blindly, challenge them and try your own, maybe better ones. Cross standards, perform experiments; explore new things instead of going where the pathways leads. Develop your own patterns, solutions. Use creativity. All that may help you not only avoid downgrade by computers but also to find a real passion and become a valuable employee even today, in the pre-AI area. We can find some nice analogy between these suggestions and the history of art in general. You have probably heard about the Impressionism art movement, characterised by open composition, clearly visible brush strokes, and its main goal being to follow emotions and attempt to capture their moments. The movement was born in the early 1860s and there were a few young painters, including Claude Monet, who introduced the style to the world. At the beginning, the style was not noticed at all and rather ridiculed by other painters, but later became widely-known, bringing Monet immortal fame and recognition (his 1872 painting entitled *Impression, Sunrise* gave the name to the art movement). Quickly, Monet gained a huge group of followers who were also talented artists, though strongly inspired by Monet. This is quite similar to the current development of AI and its applications becoming widely used tools. AI systems are able to build new pieces of solutions or solve given problems more and more beautifully with each iteration. However, they are still not able to introduce a totally new method or trend unlike anything shown to them before. AI tools are talented followers who perfectly imitate someone's style. Just like a band singing covers – it may be incredible entertainment but something is still missing... So, if you are afraid that human art or creativity is already dead, don't cross it out yet. Today's AI emulates our way of working but, as in the case of people, there is still a huge gap between follower and leader. Real creativity gives us a potential absolutely unavailable to machines at this stage and that is why it is crucial to cultivate this feature in ourselves. The less stereotypes (which are actually somehow algorithms) we follow, the more open our minds are, the higher chance to stay on the surface becoming in future not frustrated enemies but rather partners to the general artificial intelligence.

Every time I visit London, I always try to go to the National Gallery, at least for a while, to see my favourite painting, Vincent van Gogh's *Sunflowers*. Regardless of whether it is an output more of genius or maybe psychotic episodes, this illustration of what could be named a pretty trivial example of still nature (a vase with flowers) is actually widely agreed to be an absolutely world-unique masterpiece. But there is one more thing worth mentioning. One can find an ultra-high resolution (combined with millions of colours) reproduction of this painting in some online digital library, allowing study of even the smallest elements enlarged hundreds of times. But only in standing in front of the original in the National Gallery can you see some visually thicker layers of paint in particular areas of the image – and these ones are said to be van Gogh's true hallmarks. That is why no computer reproduction is able to keep its beauty, becoming just an empty file. This may be a nice metaphor for the topics we discuss – even if AI is able to emulate all of our activities one day, will it ever be a human? Rather not. There will be still some layers missing – something not necessarily easy to see at first glance but crucial to make it real.

The earlier discussion about how to protect your job from being replaced by a machine is surely an important topic. Everyone who is currently an active employee (regardless of what is his domain) needs to start thinking now about that. It also includes today's high school students who are just about to choose their profession – the application of AI covers more and more areas every single day. The process of changes on the global level has already started and surely the revolution will fully transform the labour market within the next two decades. Today the software engineer is a synonym for a well-paid job, social status and golden prey for recruitment head-hunters but a person born today may not find a job in this occupation by the time he or she would finish their studies. What is important here is that we are discussing just the next few decades. If we think of how AI may affect our civilisation and ourselves as humans within the next few generations, it will not be a topic of the workplace anymore but rather the question of human beings and their mental evolution. As we mentioned earlier in the chapter, the singularity is the moment when AI becomes conscious, creative and able to develop itself, quickly moving towards better and more advanced versions. When exactly will this moment happen? Nobody really knows but suffice to say that the variety of answers among top experts varies from just a few years to more than seven decades. One thing is certain – this moment means the end of human inventions – all research will be planned, performed and applied by AI faster then we will be able to just get this particular idea. Computers will quickly start to control all aspects of civilisation from manufacturing and agriculture, through communication and transport, to medical treatment, global governance and our everyday entertainment and activities. We can expect people in the era of the singularity will not need to work, getting all products and services on demands without any special effort or issues. You want it – you have it. Does this mean the world of our dreams? Maybe. Still, what is important is that it will also change ourselves. Just imagine we do not have really do anything. Would we be motivated enough to perform any challenging activities? I believe our future may follow one of two main paths. The first one means that people will want to learn more from machines, to extend their knowledge and skills, to be even more creative, being students or partners to the AI. Incredible facts and skills we have had no access to before may help us better understand both the world around us and ourselves. We may end up as a civilisation of philosophers and artists getting pleasure not only from products and services but even more from creative discussions, wonderful pieces of art and high level of inner peace. But we can also go in the opposite direction. If you have everything given to you, there is a temptation to do exactly nothing and enjoy laziness. We can even notice this kind of trend today. Advanced technology reduces our own skills, since they are not needed anymore. Since we are able to use GPS systems on the global scale, there are fewer and fewer people able to navigate manually, there are social problems in reading simple maps or identifying one's orientation in the world without the use of a device (by observation of the sun's movement, plant growth, etc.) and people become lost quite easily and feel states of intense anxiety when travelling without a mobile phone. The same can be seen when writing – in using computers or mobile devices to write, some people have stopped being able to write clearly by hand. We have lost many spelling and grammar skills due to the free spell checkers available in many applications (I am also using a spell checker now, when writing this book – it

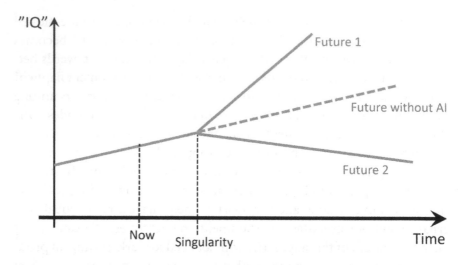

FIGURE 7.3 The future. What path will future generations choose?

has already fixed more than a dozen of my mistakes in this paragraph alone). So, unfortunately (although maybe it is something we should not worry about?), this is the direction in which civilisation could also go. We may stop using complex devices as they will be replaced by voice commands, and math skills will be one of the first to go extinct. What then? We would read and write less and less, slowly becoming illiterate. Unable to think abstractly and to understand complex ideas, we would reduce ourselves to simple consumers with narrow perspective, without ideas, creativity or dreams... If you feel upset, please remember that this may be likely, but the first scenario is also possible. And finally, in the first generation living with the singularity, it will be a personal decision which way to go. Other people may change you a lot, may limit your moves and actions, but the mind always remains free. Thus, stay open-minded and never follow the easiest path. That is the recipe for creativity and for the breakthrough inventions today, and as you can see it is also a guide for the future of mankind... (Figure 7.3).

GUNS AND ROSES

I mentioned earlier that it is much more popular amongst artificial neural network developers to fix unexpected system behaviours by presenting newer and more suitable learning examples rather than trying to manually change anything in the network structure. So, we already start to find ourselves in front of a new challenge: not only to develop a tool but also to understand what it "thinks and plans to do" (whatever these two verbs actually mean here). Understanding is the key to the future and what we need to accept is the fact that, as AI techniques become more and more complex and closer to strong AI, we will be able to mentally follow fewer and fewer of their abilities. Thinking about this for a while, we can formulate the skill levels of AI from the perspective of human abilities. Let us number these levels from 0 to 4, corresponding the future timeline:

Level 0 – we can: When we look at current AI applications, most of them are certainly now on this level. In other words, systems imitate (or emulate) some of our abilities and features, and so we have no problem understanding or repeating these automated activities. If, for any reason, an application is down, we (with our knowledge and skills) are able to perform the same actions; maybe slower, but it is still doable for us. So, for example, an OCR system used by monitoring cameras to read cars' plates – there is no problem for a human to analyse the recording manually and simply note down the numbers.

Level 1 – we can't but we know how: This level means that the AI systems are able to perform some specific actions in a way with which we are unable to compete due to our skill limitations. Still, we fully understand the application's internal structure and know how it works. We are also able to predict the rules that the AI follows. Do we have systems like that already? It might be scary at first glance, but the answer is yes. Recall AlphaGo, which I described in Chapter 3. This system defeated the human world champion in the game of Go. Some moves played by the system were recognised as extremely creative and never before played by humans (despite thousands of years of gameplay). So what does this really mean? Surely we cannot achieve the level of AlphaGo. What may be even more interesting is that we are actually unable to measure its skill level! In board games like chess, Go, etc., players are usually ranked based on match results. So, if you win against a strong player, your personal rank is increased, and after some number of successes, you may be invited to a higher division. Your rank is precisely defined as you are a part of complex mechanism based on interactions and competition. AlphaGo, being a stand-alone champion, has no one to compete with. The fascinating thing is that even if AlphaGo is further updated to improve its skills even more, there is nobody skilled enough to verify it. AlphaGo is surely just the beginning of the huge changes in front of us. There is no real problem here – the worst thing that can happen is we will simply be unable to say how strong a player it is. But if we think of future autonomous cars which are much better drivers than we are, this small problem may suddenly become a vital one. How can we check that the car is fully safe? How can we measure its skills if they are already better than ours? We may know how it works but we will be unable to verify it easily.

Level 2 – we don't know how, but we know what: This is the first of the future levels not yet reached by any existing application. It refers to systems which after a series of updates (or, even more likely, self-updates) become so complex from the code or tech perspective that we are unable to understand their mechanisms. Nowadays, although we cannot quickly explain single artificial neuron usage, we can dig for some time and by trying many options, finally investigate it – time-consuming but doable. Level 2 means that the complexity of a system and the concepts generated are unable to be followed by human developers. Imagine a very complex math proof (like Andrew Wiles' 200-page proof of Fermat's Last Theorem) – there are few people that understand its concept steps. Going further, imagine one (e.g. one found in an aliens' spacecraft) that nobody is able to understand. So, an example would be a system that optimises energy transfer so that on-the-way transmission losses are minimised which takes into consideration various parameters (including changes in requests, weather forecasts, etc.) – nobody would really understand how exactly it analyses the situation, but the action taken (saving transmitted energy) would be clear to everyone. This is Level 2, and whether we want it or not, we need to prepare for its era, maybe earlier than we initially expected. We will need to learn to trust the machines. And to make sure how we can lead their development so that they are trustworthy.

Level 3 – we don't know what, but we know why: In Level 2, even if the mechanism is too complex to be understood, we are still clear as to what the potential system does. Its activities (although too complicated to construct) remain simple from the usage (or black box) point of view. Level 3 goes even further. It refers to future complex systems for which only the overall goal will be understood. We can image a global system that controls cross-country road communications while at the same time trying to reduce air pollution, traffic jams and the number of accidents. This system would perform thousands of activities daily, some totally unclear to the users (e.g. closing a particular road or blocking a crossroad for a few minutes) mainly due to the lack of full visibility (the system would recognise potential dangers much earlier than we would). Still, the final goals which drive the AI system's behaviour are precisely defined and accepted by society. We can imagine people saying or thinking, "I do not know why this road is closed again, but I know that the system controls it well. It does a lot of good for us. There is less traffic, the air is better nowadays. So, I do not know what is happing here, but surely this must be important for making our communication even smoother, safer and more environment-friendly".

Level 4 – we do not know why: And here we reach the final skill perception level. The level that we often discuss and that evokes many emotions both in the IT industry and in the homes of ordinary people, the level that will be a simple consequence of technological evolution and, for that reason, will be inevitable. This level means that we will not be able to understand the motivation behind the AI's activities. Paraphrasing the end user sentence from Level 3, we would hear something more or less like, "I have no idea what is going on, but surely there must be a reason for this". Level 4 will be achieved as soon as AI applications attain strong intelligence, consciousness, creativity and self-motivation. The moment when AI will be able to upgrade itself and define its own goals, omitting humans in that process, we will quickly become unable to guess why a particular decision is taken. The good thing is that the AI will develop spontaneously and rapidly, not being stopped by the limitations of our minds; the risk is that it will not stay under control, and by the moment we notice any alarming symptoms it might already be too late to remove the plug from the socket…

Does this mean the future is dark and there is no hope left for humankind? Certainly not. It is quite possible that future AI (especially strong AI) will change our world into a paradise. Think of all of the deadly diseases known today with thousands of researchers looking for a cure, with only rare successes. For a strong, creative artificial intelligence equipped with a digital collection of all our knowledge (already stored in the Internet and easy to access thanks to advanced search engines – no real difficulty, with access to any resource or fact needed) might need as much as a month, or even a week, to develop a complete medical treatment for a single disease. A year may lead to the stage where all known illnesses are curable or at least stoppable in cases when the consequences (damages) cannot be reverted at some phase. The production and the preparation of international distribution may take one or two years. The dreams are just ahead. The same kind of system will help to control mass production and agriculture. Nowadays, billions of dollars are allocated to ensure food supplies in the poorest areas of the world, like many parts of Africa. Still, the money is surely not enough – millions suffer from hunger, thousands die of thirst every month. As calculated by economists, the richest 1%

of the global population owns half of the world's wealth. When money means luxury in one part of the planet, the same dollars denote survival on the other side. Artificial intelligence (at least at the beginning) will certainly not be allowed to share money equally – the rich will work hard to ensure that will not happen. But this is not the problem. For the poorest people, money is not the main need – all they think of day by day is food, water and shelter. And that is where AI can quickly help. Optimal design of land irrigation and an army of drones and droids whose responsibility is to take 24/7 care of plant cultivation, water desalination and water supply may within a decade change the world's driest continent into a place safe for its inhabitants and maybe even after another decade Africa will be green when seen from the space. Whatever is being done, human engagement is crucial, which by default means huge costs are needed to ensure salaries for people who we want to perform their activities. That is the fundamental law of economy – nobody works for free as he or she also needs money to pay other people (a shop assistant, a food factory, a landlord, a doctor, etc.). Money passes from hand to hand, creating thousands of magic circles that surround the whole planet*. There are volunteers, priests and others who do not take money at all for some of the things they do for others. That is quite an incredible attitude and we should all be grateful to them as they often gratuitously take on challenges no one else wants to. But first of all, there are just a few of them (too few unfortunately) and second, they still need money, at least for their food, shelter, transport, etc. Money certainly rules the world. But this may be changed by strong artificial intelligence. Machines work for free, they do not need food or rest. Robots can build further robots, create farms and factories that are not only self-sufficient but deliver (absolutely) free products to anyone who may need them. Money might no longer be a factor that limits helping the world. Amazing and still possible. The singularity may be the beginning of new era in our civilisation – the first time in the history when hunger, illness and poverty will be just empty words with no reference to reality...

All we have said above is a happy path, a paradise on Earth in the simplest possible description. But there is also another way that the world may proceed which is much less optimistic and much more popular in cinema. One of the most famous film directors in history, the brilliant Alfred

---

* Money sometimes even partially comes back to us with our payment – it is actually quite amazing when you try to trace the money from your wallet... You can do it yourself just for curiosity.

Hitchcock, once said that a good film should start with an earthquake and then the stress should continuously be increased. Let us be honest – a movie about a paradise with obedient robots and happy people without a trace of uncertainty would surely not attract many viewers and have no chance to compensate the production costs with ticket profits. Bad scenarios are very popular in mass media but unfortunately these are also a possibility. One of the most popular scenarios assumes that self-aware machines will try to destroy humans (for some reason). Probably most readers remember (and if you have never seen it – watch it!) the incredible *Terminator 2: Judgement Day* that I have mentioned at the beginning of this chapter. Not only its spectacular special effects but also its fresh new plot helped this picture become a classic. The Skynet (an army defence system created to identify and eliminate military risks) considers people to be enemies and launches nuclear weapons against us. Is this actually possible? As we have already discussed, even today developers are not able to fully follow all of the "impulses" in huge and complex artificial neural networks – in other words, we cannot describe all of an AI's thoughts, so we are unable to fully predict all of the consequences and actions performed by machines. And although the probability of a mistake (or incorrect behaviour) is much less than in the case of humans, we need to remember that there are still other "blockers" that we have in our minds (like empathy, conscience, beliefs) which a computer is free of. If your calculated result (even that of a high value) suggests the need to attack somebody (or some other country), you will surely review the decision once again yourself. World presidents and other decision makers are surrounded by armies of analysts and advisers, and any military decision is carefully considered before it is made. Automatic decisions of such significance may lead to some unexpected and dangerous situations. Moreover, if you teach a machine to identify enemies, there is always a chance it will point at you at some stage. So, the areas of AI application are crucial. Using AI in the military industry increases the risk of Armageddon. That is why during the opening of one of the world's biggest AI conferences in 2015 – the International Joint Conference on Artificial Intelligence (**IJCAI**), an open letter was announced to request a ban on offensive autonomous weapons that are beyond meaningful human control. More than 20,000 people signed the letter including people as famous as Stuart Russell, Stephen Hawking, Elon Musk, Noam Chomsky and many more. Unfortunately, the letter was, at least partially, ignored. In the same year, the US Army announced the success of their first series of tests of armed military swarms of dozens of drones attacking a target together, communicating with each other and co-operating without human supervision.

In 1999, the Wachowski siblings brought to the big screen another new concept. *The Matrix* presents a seemingly normal world like the one we all live in. But the truth is not visible to the eyes of the majority: everything we perceive is a computer illusion, a complex virtual reality called the Matrix. So, what is the truth? People are cultivated by intelligent machines and stored in small tubes filled with nutritious liquids – all to collect electric energy produced by our brains. Humans changed into batteries – is this our future? The recent incredible progress in virtual reality technology surely does not allow us to cross out such a possibility. What is more, it may be something desired by many in future generations – the chance to live in an emulated and fully safe paradise where you cannot be hurt or encounter any kind of discomfort. That could be surprising and unacceptable at the very first glance until we think about this much deeper: a virtual world that we control and design ourselves might be one of the steps of the evolution of civilisation. It is widely discussed that the future of world exploration may follow one of two main pathways. Either we will continue to explore the physical universe, digging and diving deep under the Earth's surface and building rockets and measuring devices to move even further the edges of the known Universe. But the other option is that we will focus on the virtual world and decide to easily create our own universes instead of spending tonnes of energy to discover the real one. In the 1970s, the famous **Fermi paradox** was formulated: if there are millions of billions of stars and thus an uncountable number of potential civilisations across the Universe, how it is possible we have not encountered any evidence of their existence? Where are they? Is it possible we are alone in so huge a space? Or maybe they are there but do not wish to answer our calls for some reason? Maybe at some advanced staged of any civilisation (advanced enough that they could reply) they decide to ignore the surrounding world and focus on the virtual one, staying on their planet in full silence... Of course, we need to consider the more pessimistic variant too. Maybe achieving the singularity is the moment that any civilisation ends. Is AI the final stage of evolution, a mystery mechanism designed by nature to eliminate any species that dominates it too much?

The imprisonment direction shown in *The Matrix* may not be the only one. Cinema usually presents us with malicious, evil machines that close us in cages or try to exterminate us like vermin. But if you think of the Matrix a little bit deeper, you can realise that most of the population (actually all except Neo and his colleagues) live quite a normal live, with an average salary, various passions, enjoying their families and some free time too. They were not physically hurt and have no idea about their imprisonment. Still, is it always your enemy who limits your activities and free will? Certainly not. Parents often do not let their kids do exactly what they want to – they try to protect them, knowing that their young minds are too little-experienced to predict (and avoid) all the danger that the world is filled with. If you hear some memories by the members of royal families, the phrase *golden cage* is one that occurs in almost all interviews – these kids are often not allowed to behave spontaneously or do anything not following the rules and protocols (which are often not even up-to-date). So, one negative scenario we can imagine is one in which the AI of tomorrow will treat us all as valuable treasures, which may seem correct at first glance. But being overprotective may not be what we really want of our machines. The system may treat us like small kids, not allowing us to perform any activities (to make sure we are not hurt) – robots could block us from doing sport (the chance of concussion is quite high), eat whatever we want (including tasty but greasy burgers) or relax in a form we prefer (no beer, not too much sun or sex). At some stage we may end up being closed into our homes as the world outside is too dangerous – healthy and long-lived but without any chance to find what a full life is all about.

There is actually one more (of many) possible bad end for us when looking at the possible implications of the singularity. Let us suppose that AI will control most of our world one day – something the current development and progress moves towards. It will control transport, energy, agriculture and farming, even weather – all to make our own lives simpler, cheaper and more efficient. All that sounds perfectly fine. But remember what we said in Chapter 2: something that is simple for humans is extremely difficult for machines and vice versa. The more "intelligent" systems become the more human-like they are. They work faster and more in the way that we would but there is one more feature they can inherit together with all the expected ones: the chances of making a mistake. A standard computer calculates numbers – there is no space for a mistake there. But with the increasing complexity of applications and problems being analysed, mistakes could start to occur. This is not surprising – even a human genius makes mistakes. The only problem is if the future system controls almost all the world, a single mistake may imply dramatic consequences. What if the mistake refers to food or production of a new, brilliant vitamin– in the worst case it may mean that most of the world's population would die of hunger or for example become sterile (so we would simply go extinct, having no descendants). The more responsibility we give to machines, the higher risk we need to consider. Will our logical thinking and foresight defeat built-in laziness?

During my lectures or discussions, I quite often hear the following question: there is a lot said nowadays about the future, happy paths and traps that may occur on our progress as a civilisation, but is this really going to happen? Will AI really dominate the world of tomorrow or is it just a nice and trendy pop topic? Is there any reason to believe it? Yes, there is. The reason is that it is already happening as we speak (or as you read this book). It is enough to open your eyes wider, to look around more carefully and to focus on the signs of how important AI has started to become. We have already mentioned the open letter by scientists in which they postulate a legal regulation to ban the usage of AI in autonomous weapons. The timing is not accidental here – many of the signatories had a view on the progress and research being done by the military industry. Their warning is not science fiction or self-promotion to reach a wider audience – many of them are already rich and well-known. So, it is not for fun or entertainment – it is to protect our world as a home for all of us. We might even now realise how close we already are to the breakthrough moment. In the same year as the open letter, the famous **Bilderberg Group**, an annual, prestigious and quite confident meeting of world's most powerful or influential politicians, economists and businesspeople, discussed the topics of European strategy, globalisation, the Middle East, NATO, terrorism... and artificial intelligence. So, if they spent time on talks around AI, we should definitely start to think of this much more seriously than we have done before. Again, the same year, 2015, **OpenAI**® was founded as non-profit artificial intelligence research company to develop safe solutions in AI and ensure the results are shared as widely and as equally as possible. The input, estimated to be an incredible 1 billion dollars, gives the clear message that this is surely not a playground. Private people investing such money must know it is worth doing so. Finally, in September 2016, the **Partnership on AI** was announced, bringing together the world's IT giants including (but not limited to) Amazon, Facebook®, Apple®, DeepMind and Google, Microsoft®, and IBM. The unified power of these companies is, officially, focused on formulating best practices (standards) on AI technologies, to serve as platforms for discussion, knowledge exchange, and monitoring AI influence on people, societies and the whole world. Officially. But it is sometimes worth recalling an old Chinese saying, traditionally attributed to Sun Tzu, the 6th century BC military strategist: *keep your friends close and* *your enemies even closer.* AI is a growing area of business with limitations that are difficult to actually point out. There is surely a second purpose to the partnership – to be closer to the competition, to have more chances to monitor one another's progress and results. Whatever is said officially and whatever we may guess or speculate here, one thing is certain: the first company to build a conscious system is the one to win all the pools. There will be no runners up or consolation prizes. Even the giants are fighting for survival here as artificial intelligence will dominate IT quite soon. Anyone not ready to jump on the train on time will not be able to catch up and will simply disappear from the market. AI is probably the biggest opportunity for business and civilisation in history but it will be a single moment situation – when the cards will be dealt with no chance for a rematch.

Just two final thoughts in this section. The first one – the list of absentees is often equally as important as the list of participants – I will leave to the reader to think for themselves as to why some companies have not joined various licences, standards or partnerships like the ones described above. Not feeling comfortable? Not seeing the point? Or maybe you are too close to the treasure to want to share the map… The second thing – most of the above examples come from 2015–16. Progress since then has not frozen but has even sped up. Check the date in your calendar and multiply your expectations, feelings and predictions by three…

TABULA RASA

Whenever I am asked about the evil and bloodthirsty machines of tomorrow I like, in reply, to recall the philosophical concept of the **tabula rasa**, traceable to the writings of Aristotle – an ancient genius and the one of the fathers of science as a whole. The phrase *tabula rasa* comes from Latin and means "blank slate". It refers to an idea that people are born without any built-in, mystic or extraordinary mental content – we appear in this world with our brains as empty as a brand new pen drive bought in a local tech store. All we know, all we are, is due to the perceptions and experiences collected during our lifetime. This concept has as many enthusiasts as opponents as it partially touches on the always-controversial topics of souls, religious beliefs, reincarnation and more. I do not want to start such a discussion here – I never try to push on any of the options, leaving it fully to the interlocutors – our beliefs are important parts of our lives, and if we want to change them, we are changing the person himself or herself. However, I have recalled the *tabula rasa* concept as in my opinion, it nicely describes artificial intelligence. It is clear that machines and AI systems are "blank slates" at the very beginning. People who think about machines being born bad are wrong – even the most complex artificial neural network has some default set of values assigned to its internal connection weights. There are no hidden built-in instincts, feelings or prejudices. It is important to understand that AI by itself is by definition neither good nor bad. It is a tool, like a hammer – which we can use it to build a home or to hurt each other. The decision belongs to the user. And that is where the real danger is – it is not the machine which is a risk factor, but rather its developer, its architect – a human in general. We, and nobody else, have the power to change it into a helpful tool or to a weapon. Thus, keeping this in mind, it is crucial to choose appropriate applications and precise goals for a machine (which it will be dedicated to follow), as well as working together on legal regulations. If we create a military AI system to search for and eliminate our personal enemies, there is a risk that the algorithm (while learning constantly based on the examples, collecting experience) may one day identify all humans as enemies… and decide itself to eliminate us – paradoxically, precisely following our own rules and the goals we ourselves put into it. On the other hand, there is no chance that AI will want to destroy us if its creators have not defined a concept of *enemy* at all (Figure 7.4).

FIGURE 7.4 A hammer – a tool or a weapon?

Choosing a safe application is just one of the things that must be considered. We also need to remember that any AI system (especially those based on neural networks or genetic algorithms) constantly improves day by day by encountering new cases and learning by example. It is not obvious at first glance but it is rather certain that some periodic examinations have to be performed to check exactly how the system is working. We can imagine a clear analogy with plane pilots, taxi drivers, soldiers, etc. They are well trained experts, high level professionals. However, during their lifetime there are many factors that influence their physical and mental condition, e.g. a traumatic event might irreversibly affect one's character and perception of the world. That is why such experts (especially the ones whose work, if done incorrectly, can threaten somebody's life) have periodic obligatory tests. AI must be subjected to the same rigorous, periodic tests.

Finally, it is crucial to understand that every single message or input changes the neural network structure. At the same time, we need to highlight that future AI systems will not be isolated but rather will be designed to interact with other applications and hundreds of people. We must somehow ensure (by internal blockades) that the external environment won't influence AI behaviour in an unexpected way. In an age when almost the entire population has Internet access, and even young people knowns how to explore it in all directions, in the era of viruses and hackers, this may be the biggest challenge and the biggest risk of the whole topic of AI.

✎ NOTES

- The singularity is the future moment when strong artificial intelligence is created. From that point we can expect exponential growth in science and IT development. Strong AI will be our last and biggest invention, and all further discoveries will be done by machines.
- The more complex computer systems become, the less we will know about their internal low-level mechanisms. The same happens with the services and good that surround us – we do not really know (and do not care) how they are provided or produced. Each new solution is more general, more precise or simpler. The black box (the area of unknown detail) constantly increases.
- Progress in AI will sure influence the labour market. Some job positions will be reduced or even eliminated in future, while others (currently undervalued) might be found much more valuable than nowadays.
- To keep your employment and not be replaced by machines, do not behave like a machine: do not act automatically, avoid work according

only to scripts and checklists, break stereotypes, cross standards, develop your own patterns, use creativity, follow intuition.

- The future controlled by machines may open two pathways for human-kind – either we will become illiterates focused on the simplest pleasures or we will try to learn from machines to become a society of artists, philosophers and scientists. It is up to us which road we choose.
- There are both happy and scary scenarios for the future after the era of the singularity era begins. Still, it is important to highlight that AI is nei-ther good nor bad by default. Like a *tabula rasa*, it is an empty system at the very beginning. It is fully our own responsibility what we fill it with. AI is a tool like a hammer – which you can use as a tool to build a house, but can also use as a deadly weapon. We have the keys to change the world into a heaven or a hell. So, the human factor is crucial here.
- Finally, a future controlled by AI is absolutely certain. The question of *if*, still being asked maybe a decade ago is no longer valid. The question repeated today is *when*.

✎ **YOUR NOTES**

Final Thoughts

D URING MY PRESENTATIONS AND talks for a wider audience, I am very
often asked about the future of artificial intelligence and the chances
for Hollywood scripts to one day become reality. Movies are by definition
characterised by breathtaking visualisation, thrilling plots and controver-
sial theses – otherwise the whole cinema industry would fail due to the
lack of viewers. But is it possible that a Skynet-like system would take control of US nuclear weapon and aim missiles at the world's biggest cities? Or
that computers would decide to breed us as natural batteries, as shown in
the *Matrix* movies? Myself, I am rather optimistic on this. Machines are
devoid of instincts and hidden thoughts – their whole source code starts
from an empty file and thus, as long as they are developed carefully, computer systems should have no reason to harm us. Of course, the future is a
mystery and we can never be sure about the way in which AI will evolve.
Still, I am much more afraid of humans than machines. Why? Imagine a
world fully controlled by strong AI, a world without illnesses, hunger or
poverty, where everyone has his own house and car and does not want
for anything. A global nation, without money (no longer), where all are
equal and free to experience a life full of happiness. Sounds like a paradise,
doesn't it? But are we mentally ready to live in paradise? For millions of
years of evolution, we have been being prepared to fight for survival, to
protect our territory, to stay careful and mistrustful. Our instincts, coded
inside the (historically) oldest parts of our brain, still drive our actions
when in stress or fear, blocking advanced intellectual abilities. That is scientifically proven. If you do not believe in that, think about your reaction
to a sudden loud noise (like fireworks exploding nearby) or recall images

of chaotic crowds in panic mode. Evolution prepared us well. Ambition, motivation, rivalry – all these nowadays help us to overcome the adversity of fate. They help both a young businessman climbing a corporation's ladder (to gain power and influence in exactly the same way as joining the council of a tribe) and a seriously ill patient to recover after a near-death experience. We spend a lot of energy just to keep going, on our usual challenges, responsibilities or problems. So where would we allocate it when having everything given to us by default? Will unlimited free time lead us to creativity and art, or rather to aggression and the call of anarchy? Maybe it is better for our society to have everyone busy with everyday matters? What has taken millions of years for evolution to code in our brains may be difficult to change within a decade. We may look for enemies instinctively, even if there are no real opponents. Think about people working hard for years to build a house while their lazy neighbours get the same constructed by machines? Or royal families used to ruling and making global decisions? And, on the other hand, people who are just happy to be told what to do and managed by governments. The consequences of the sudden disappearance of social divisions and traditional roles may be difficult to predict. Surely not all would immediately accept and understand such a change.

The second question I often hear is the one about what the future, strong artificial intelligence could possibly be. Will it replace us? Will machines one day become exactly the same as humans, with no possibility of distinguishing one from another? These questions are very difficult to answer, as, when we think deeper, we quickly notice that the answers may be found in philosophy, ethics and religion rather than only in computer science. The source of all behaviours and actions performed by a machine is written in the source code, nothing more and nothing less. Due to this, the whole artificial brain and consciousness can be simply copied from one device to another, quickly transferred if in the case of issues with the initial hardware. This makes the AI immortal. But maybe that is not something we should be jealous of? Mortality, like growth, is one of the essential components of life. We can philosophically say that if you are not designed to die you do not really live. Fear of death somehow makes us human. *Memento mori** motivates us to value quality of life, to

---

* Latin phrase: "*Remember that you must die*", which reflects the way of thinking in Medieval times. It was the ultimate message of the Christian religion and also the fascination with death which united all people independent of their social status (fr. *Danse Macabre*).

care about our families and societies as being a form of continuity of our limited existence. It makes us interested in history and our origins and drive us to create art as a memorial for future generations.

We discuss whether computers will become like humans but the question could also be reversed: are we becoming more and more like machines? Small mistakes, funny misunderstandings, clumsiness – we often find these as our weaknesses, but the truth is that these little failures enrich our stable life, adding some colour to the everyday monotony. In today's world, where everything happens in a hurry, by rules and by standards, people surrounded by technology behave in a specific, society-agreed way. Many say that their life is boring, too arranged, with everything planned carefully from morning till evening. They feel like robots in the work they spend most of their time at. So, do machines become more and more similar to us or is it maybe our actions which are becoming similar to software programmes? What about the diversity of thoughts and ideas among people – is this slowly disappearing? How about our spontaneity – do we still have at least some of it? Nowadays we are used to calculating everything, money, chances, probabilities. Exactly like artificial intelligence. And maybe the most important question: are we still able to selflessly love, to make an **uncalculated** personal sacrifice for the good of our societies or a stranger?

The future is a mystery. But one thing is certain – we need to care more about the features that make us human. And there is one more paradox on the horizon: in studying artificial intelligence, we may soon learn much more about ourselves than about computers.

YOUR THOUGHTS

Index

Activation function, *see* Transfer function
Adelson, Edward, 60
Alexa® (Amazon), 123
Algorithm
    and heuristics, difference between, 15
    working and significance of, 16–18
AlphaGo®, 34–35, 75–76, 151
Aquarium metaphor, 37, 38
Archimedes, 113, 114
Aristotle, 129, 131, 142, 159
Artificial art, 101
Artificial brains, 49–57
Artificial creativity, 101
Artificial DNA, 86–87
Artificial reasoning, 64–74
Automated image recognition, 20
Automated translation, 123

Backpropagation, 68, 69, 71
Base, 47
Big picture, perception of, 32–33, 73
Bilderberg Group, 158
Binary system, 46
Black box, 52, 72, 141–145
Bombe calculating machine, 49, 143
Brainstorming, 57–61
Brute force, 30

CAPTCHA test, 45
Ceulen, Ludolph van, 114
Chabris, Christopher, 58
Checker illusion, 60
Chomsky, Noam, 155
Chromosome, 87
Circumscribed square, 116
Cloud computing, 75
Collocations, 129

Communication, significance of, 135
Consciousness, 10
Context, 133
Creativity, 9
Crossover, 91–96
    of coins chromosome, 100

Darwin, Charles, 82
Deep Blue supercomputer, 29
Deep learning, 74–78
Dice, 111, 115–116

Emulation, 8
Error rate, 70, 71
    illustration of, 72
Evolution, 83–86
    genetic algorithms and, 101–102
    of solution, 97–101

Fan Hui, 34
Fear, significance of, 2
Feelings, 11–12
Fermi paradox, 156
Fitness function, 90–91, 93
Frank, Andrew, 37
Fuzzy sets, 21

Games, 25–29
    chess, 26–29
    Go, 30–35
    Sudoku, 66–68
Genetic algorithms, 23, 81–86
    artificial DNA and, 86–87
    crossover, 91–96
    cycle of, 87
    evolution in, 101–102
    life birth and, 88–89

mutations and, 96–97
natural selection and, 89–91
solution evolution, 97–101
Genus–differentia definition, 131
GFLOPS, 29
Go game, 30–32
Google® DeepMind, 34
Gradient descent optimisation
algorithms, 73

Hawking, Stephen, 155
Helicopter view, *see* Big picture,
perception of
Heuristics, 16–17, 24
and algorithms, difference between, 15
importance of, 19–23
meaning and significance of, 19
Hexadecimal system, 48–49
Hitchcock, Alfred, 154–155
Home® (Google), 123
Hyponymy, 132

IJCAI, *see* International Joint Conference
on Artificial Intelligence (IJCAI)
Independent events, 111
Intelligent information bypass, 125
International Joint Conference on
Artificial Intelligence
(IJCAI), 155
Iterations, 55

Kasparov, Garry, 29
Kelvin, Lord, 140
Keywords technique, 130
Knapsack problem, 22–24, 101
Knowledge, 36

Language processing, 121–126, 134–136
semantics and, 128–134
syntax and, 126–128
Layers, 57, 61–64
hidden, 62
input, 61–62
output, 63–64, 70
Learning set, 54–56
Lee Sedol, 34, 35
Let's Make a Deal®, 111–112
Life birth, 88–89

Linear and exponential growth, difference
between, 140–141
Lovelace, Ada, 49–50
Ludolphine number, *see* Pi

Maddrell, Ned, 135
Mark test, 11
Matrix, The (film), 156
McLemore, Doris, 135
Meronymy, 132
Millennium Prize Problems, 25
Mirror test, 11
Monet, Claude, 147
Monte Carlo method, 105–112
pi and, 112–118
Monty Hall problem, 111
Musk, Elon, 155
Mutations, 96–97

Natural language, 123
Natural language processing (NLP),
123–124, 136
syntax and, 127
Natural selection, 89–91
Network topology, 62
Neural networks, 41–47
artificial brains and, 49–57
artificial reasoning and, 64–74
brainstorming and, 57–61
deep learning and, 74–78
layers, 61–64
positional systems and, 47–49
Neurons, 42–43
artificial, 51–52
NLP, *see* Natural language processing (NLP)

Ontology, 131, 132
OpenAI®, 158
Overfitting, 54

P = NP problem, 24–25
Partnership on AI, 158
Pattern, 43–44
Pi, 112–118
Plato, 131
Plato's Cave, 131
Polint-112-SMS prototype, 125
Popper, Karl, 143

Population, 89
Positional systems, 47–49
Probability theory, 110–111
Pseudorandom numbers, 108

Random number, 107–110
Reverse Turing test, *see* CAPTCHA test
Rucksack problem, *see* Knapsack problem
Russell, Stuart, 155

Seed, 108
Selective attention, 58
Self-awareness, 11
Sentiment analysis, 134
Sexagesimal system, 49
Shaw, George Bernard, 96
Simons, Daniel, 58
Simulated annealing, 73–74
Singularity, 140, 141, 149, 150, 154
Skill levels, of artificial intelligence, 151
Smith, Fanny Cochrane, 134–135
Squaring, of circle, 112
Strength, 35–36

Strong artificial intelligence, 8–12
Sub-languages, 133–134
Sudoku, 66
Sun Tzu, 158
Synapses, 43
    identification of, 77
Synonym ring, *see* Synset
Synset, 132

Tabula rasa, 159–161
Terminator 2: (film), 139, 155
Testing set, 54, 56
Topology, 72
Transfer function, 51, 68
Turing test, 12–14, 45

Watson, Thomas, 141
Weak artificial intelligence, 8, 14, 21, 35
Wedding pyramid, 43
Weight, 57
Wiles, Andrew, 142
Wisdom, 36
Writing, invention of, 121–122

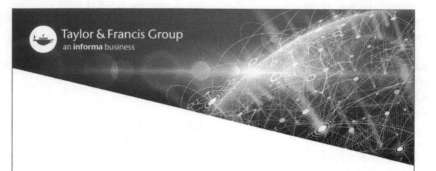

Milton Keynes UK
Ingram Content Group UK Ltd.
UKHW040054071024
449327UK00019B/551

9 780367 898021